Can Science and Religion Live Together?

What Science Can and Cannot Do

by

Gerard M. Verschuuren

En Route Books and Media, LLC
St. Louis, MO

✸ENROUTE

Make the time

En Route Books and Media, LLC
5705 Rhodes Avenue
St. Louis, MO 63109

Cover credit: TJ Burdick

LCCN: 2020941667
ISBN-13: 978-1-952464-17-1

Table of Contents

Preface i

I. An X-Ray of Science..1
   1. A Simplistic View... 2
   2. What's Wrong with the Simplistic View?................7
   3. A More Realistic View 25
   4. Ten Examples from the History of Science........... 36

II. Science, and Only Science?.. 65
   5. The Megalomania of Some Scientists 66
   6. What's Wrong with Scientism? 76
   7. Science Is Never the Final Truth........................ 84

III. How To Get the Facts?...91
   8. What Are Facts? 92
   9. From Concepts to Facts and Back.......................103
   10. Can the Facts Change? 110

IV. Where Do Facts Come from?.................................... 121
   11. A World of Concepts...122
   12. A World of Propositions 128
   13. The Divine Intellect 133

V. No Science Without Religion......................................139
    14. The Assumptions of Science............................. 140
    15. Assumptions from Heaven.................................155
    16. The Judeo-Christian Roots of Science.................162

VI. Religion and Science Need Each Other 171
    17. The Author of Two Books...................................172
    18. God and the Big Bang...185
    19. God and Evolution..198

VII. Conclusion 211

For Further Reading..217

Index 219

Praise for the Book 225

About the Author.. 227

Preface

Undoubtedly, science has one of the most impressive track records in human history. It is a success story that keeps persistently adding new achievements to the list, with no end in sight. So it shouldn't come as a surprise that science has given us reason not only for high hopes, but also for extravagant claims.

To find out what those hopes and claims are worth, this book begins with an explanation of what science can indeed do for us and how. Most people don't need to be convinced, though, that the power of science is enormous. Just look around to see the achievements we owe to science: the Big Bang theory, space exploration, the human genome project, antibiotics, vaccines, cancer treatment, and the list goes on and on. We could not live the way we do nowadays without the fruits of science.

But this book is also about what science is *not* able to do for us. I don't mean those things that science is still searching for, but rather those things that science has no access to whatsoever—things such as thoughts, values, beliefs, hopes, dreams, and ideals. Are all of these merely the result of material, molecular interactions? Isn't love more than a chemical reaction, and aren't thoughts more than brain waves? And most of all, what about religion? People who think that science has no limitations whatsoever—it's just a matter of time, in their view—should think a bit longer and a bit deeper. This book will help them to do so and will come up with some startling conclusions.

Science can never discover its own limitations, for that requires philosophy. I realize very well that most scientists feel uneasy with philosophy, some actually hate it. "It has often been said, and certainly not without justification, that the man of science is a poor philosopher"—Albert Einstein said it, not I. Indeed, philosophy may seem a pretty sophisticated enterprise, but there is probably not one person who doesn't philosophize. Each one of us is destined to start philosophizing at some point in life, but some better than others. One of the tasks good philosophers have is questioning and analyzing what most other people, including scientists, usually take for granted. When philosophers look at science, they not only see what science is able to do for us, but also what science is *not* able to do for us, and actually cannot possibly do for us.

That's where the door for religion has opened. Really? Hasn't science shown us there is no God, you might say. The big surprise of this book will be that a serious philosophical reflection of what science can do for us actually will reveal to us that science cannot operate properly without God and without some kind of religion.

That's probably hard for many to believe. The fact of the matter is that there seems to be a real tension, if not a serious conflict, between science and religion. Here is one of the reasons: Science is making progress, but religion does not seem to. Science discovers more and more about nature, but religion no longer seems to discover more new things about God. No wonder this has made many people suspicious of religion.

The Catholic philosopher Peter Kreeft compares the situation with that popular Western in which one cowboy says to the other: "This town ain't big enough for both of us. One of us has to leave." Many think of science and religion in a similar way: there is not enough space for both science and religion in this world—one of them has to leave. If one of the two makes progress while the other does not, then the one who gains must do so at the cost of the other. But is that really true? More than ever do we need philosophy. That's why we need this book, to show us that the world is "big" enough for both science and religion. On further philosophical inspection, there appears to be so much that science is "blind" to. Science may be everywhere, but science is certainly not all there is.

Let me make clear from the outset that when speaking about science in this book, I am referring more specifically to the so-called natural sciences—astronomy, physics, chemistry, and biology. And when speaking about religion, my focus is primarily on the Judeo-Christian religion—more specifically, the Catholic religion. Let's find out how they can live together—or actually cannot live without each other.

I

An X-Ray of Science

In most of the Western world, and even beyond, scientific research is widely seen as a highly esteemed activity. Science is supposed to be the final judge and referee of all we know about the world around us. We like to conclude any discussion with the statement, "Science has spoken." Even commercials claim "scientific evidence" that one product is better than another.

Science has acquired enormous authority. It is often associated with "certain" and "proven" knowledge. Indeed, the word "science" is related to the Latin word for "knowing" (*scire*); and in most philosophical traditions "knowing" is identical with "knowing for sure." On the surface, such a criterion seems to create a clear borderline between "scientific" and "unscientific" statements.

However, there is science, and then there is science. What science really is like is hard to define. There is much disagreement, even among scientists themselves. Most agree that physics and chemistry are "real" sciences, often collectively referred to as "hard" sciences. But what about psycho-analysis, to just mention one case of disagreement? Even certain parts of the life sciences such as paleontology are questioned by some as to whether they deserve to be called science. Apparently, there is a great deal of debate as to what science entails. Yet, all sciences seem to share some common methodology of testing their predictions by observations and/or experiments.

As a matter of fact, scientists explore their field of study most of the time in a way we would call *observational* and *experimental*—which means that they "use their (aided) senses" to gather data and they "use their (armed) hands" to manipulate their objects, and more recently they also "use their computers" to simulate their models. The main difference between an observational and experimental approach is that the latter approach lets the scientists choose the conditions of the experiment, which they cannot in the observational approach.

So most people take science as a systematic enterprise based on empirical observations, experimental techniques, and computer simulations. It is an enterprise which creates, builds, and organizes knowledge in the form of testable explanations and predictions about the Universe—ranging from planets, atoms, cells, diseases, brains, to dinosaurs. No wonder, scientific research seems to be a game for insiders—that is to say, a sophisticated game. Anyone who wants to join the game has to be initiated into the rules of the game; from then on, the rules have to be mastered and applied. What rules are they?

1. A Simplistic View

A simplistic view of science has it that "real" knowledge—that is, scientific knowledge—is only possible as a result of measuring and counting. An inscription on the façade of the Social Science Research Building at the University of Chicago reads, "If you cannot measure, your knowledge is meagre and unsatisfactory."

How do we get to such allegedly secure scientific knowledge? The answer is quite straightforward: by using our senses and making observations. It is by using (aided) senses for observation, (armed) hands for manipulation, and (computerized) tools for simulations that scientists collect "observations." Once they have repeated the same observation multiple times and have a series of singular observation statements, they can generalize them into a universal law—for instance, of the type "All metals expand when heated," or "All solids dissolved in a liquid raise its boiling point," or "All radioactive elements decay at a certain rate" or "All acids turn litmus paper red," or "All organisms are exposed to natural selection," or "All organisms require a source of energy," or "All organisms contain one or more cells." According to the simplistic view of science, the keyword should be "all."

What the scientists did in the previous cases is making the step from "some" to "many" cases, and ultimately to "all" similar cases—that is, from a *finite* number of cases to an *infinite* number of cases. It is the decisive step from singular observation statements to the statement of a general and universal law. Why is a singular observation statement not enough? One of the reasons is the fear of "anecdotal" or "circumstantial" evidence that relies on a single observation, or on only a few of them. Instead, the more similar cases scientists have, the more certain their knowledge becomes, proportional to the number of observations supporting it. This process is called *induction*. People who believe that induction is the favorite, if not only, method of practicing science are called *inductivists* (sometimes equivalent to empiricists or positivists). They believe induction makes it possible for science to grow.

The inductivist approach of science portrays science as an accumulation of singular, empirical observations of a particular connection—for example, the link between acids and litmus. Finding a link between two variables is often considered to be the main goal of scientists, at least to begin with. Very often, scientific research starts out with a simple model of, say, two variables. The independent variable is the one freely chosen by the scientist in order to trace its influence on the other, so-called dependent, variable. With the help of these terms, we are able to define a scientific research "problem" in a more specific way as follows: what is the influence of the independent variable on the dependent variable?

The problem is, however, that apart from the two variables studied in an experiment, there is an unlimited group of so-called non-experimental variables which may potentially interfere with the relationship between the dependent and the independent variable. The nightmare of all scientists is that their research is going to be blown by some non-experimental variable. Hence, it is of great importance to eliminate the influence of these non-experimental variables in advance and to the greatest degree possible.

Another problem is: how do we know the supposed link between two variables is reliable, in the sense that the independent variable really *causes* the dependent variable? We don't want to end up with merely "anecdotal" or "circum-stantial" evidence based on only a very few "haphazard" obser-vations. That's the main reason why inductivists do aim for a *series* of repeated, identical observations. For instance, if dipping blue litmus paper in acid is repeatedly accompanied by the observation of the paper's turning red, we may conclude that dipping blue litmus paper in acid is the *cause* of its turning red. This is a conclusion about *all* similar cases based on *some*, or preferably *many*, similar cases. That's the way induction works.

Is there any logic behind induction? And if so, how safe is this kind of logic? It is basically a form of "generalizing induc-tion," which takes us from a general statement about *some* observations to a universal statement about *all* observations. One starts with singular statements about observations, and then by adding more and more similar observations, one comes closer and closer to a universal observation statement. After having seen many times that iron expands by heating, one concludes that all iron does so. That is a way of doing research that seems to have worked very well in science over and over again. But ironically, that argument in itself is a form of induction used to justify induction.

However, we do not have here a logically safe kind of reasoning. Induction is different from deduction. Induction teaches us new things we didn't know yet, whereas deduction teaches us only things we knew already but may not have realized we knew them. Deduction is a logically safe form of reasoning, but induction is not. Take, for instance, the conditional premise "If X holds for all iron, then X holds for some iron." If there is a second premise affirming that X does hold for *all* iron, we must conclude that X holds also for *some* iron. This would be a *deductive* argument, a safe form of reasoning. But that's not the typical form of reasoning in science.

The previous argument would be *inductive* if the second premise says that X holds for *some* iron. In science, we usually conclude from this that X holds for *all* iron. But, unfortunately, that is not necessarily the case. Let's consider another comparison. Deduction is a form of reasoning from "the top down": all organisms require a source of energy; well, I am an organism; so I require a source of energy. That's deduction. But induction is a form of reasoning from "the bottom up": I require a source of energy; well, I am an organism; so all organisms require a source of energy. The former form is logically valid, but the latter is not.

In other words, an inductive argument is basically a logical fallacy. To conclude, based on repeated observations of a connection between two variables, that the independent variable (X) is the *cause* of the dependent variable (Y) is not logically warranted. Take the well-known saying "After this (X), therefore because of this (X)." It is an invalid argument based on the legitimate assumption that all events have their cause in what has gone before they happened. In other words, "because of X" implies "after X"—well, if Y happens "because of X," it must be "after X." But the opposite derivation is not valid. If Y happens "after X," it *may* be "because of X." Since induction is not a safe kind of logic, inductivists require many repeated observations. In essence, inductivism is a way of building theories up through induction—that is, "from the bottom up," from case to case, from observation to observation, from general observations to universal observations.

Of course, even inductivists of the simplistic view are aware of the old, classical warning that says: "Post hoc sed non propter hoc"—in translation: "after X" and yet not "because of X." For instance, you got better (Y) after taking some medication (X), but was getting better (Y) also because of taking some medication (X)? Was X really the *cause* of Y? Perhaps so, but not by the force of logic, and therefore not necessarily so. That's one of the reasons why scientists make an important distinction between causation and correlation. One of the first things you learn in a science class is that correlation doesn't imply causation.

Are there ways to make the case for causality stronger than mere correlation? Inductivists of the simplistic view think so. The classical "heroes" of induction are Francis Bacon and John Stuart Mill. The former believed he had found a method stripped of all philosophical fancies. The latter believed he could offer us an explicit set of clean and safe inductive rules which would help us in a "mechanical" way to seek and find the *cause* of something in particular—"from the bottom up," so to speak.

The first of Mill's rules, for instance, pretends to tell us exactly how to find the cause of something. Its recipe is as follows: look for all the circumstances preceding some phenomenon and find out which circumstance in particular occurs every time the phenomenon takes place. In finding that particular circumstance, we have demonstrated by induction that in the many cases studied, there is one particular circumstance, the independent variable, which is the *cause* of the phenomenon in question. Thus, we would end up with a universal statement to the effect that circumstance X is the cause of phenomenon Y in *all* cases—observed and unobserved. If a certain sickness, for instance, occurs in human beings who all carry a certain type of bacteria, we must assume that it is this kind of bacteria that causes the sickness.

Mill gave the inductivists several more rules of doing research. But the bottom-line is this. According to the inductivist model, a scientist observes nature, then tentatively poses a modest statement to generalize an observed pattern, then confirms it by more repeated observations of the same pattern, next ventures a modestly broader statement, and finally confirms this as well by adding more and more observations. But the fact remains that the inductivist approach builds science on a very weak basis—a rather limited conception of observation. It is only through observation statements that science is supposed to grow according to the simplistic view.

2. What's Wrong with the Simplistic View?

Why does inductivism fail us when we want to explain how science works and what science can do? Studying the effect of one variable on another variable through repeated observations may sound very nice and simple, but even with such simple cases, there are several problems inductivists need to face. We will discuss at least ten of them here.

(1) Probably the most central assumption of inductivism is a fundamental disconnect between the collection of observation statements and the body of scientific theories. In this view, it is the former that determines which theories are to be developed and accepted. The system of postulates which make up a theory supposedly "floats" or "hovers" freely above the plateau of observation statements. In this view, theories emerge nicely from observations and find their justification in observations. This creates the false impression that the objectivity of science is guaranteed by the "pure empirical facts" found "at the bottom," which are supposedly known regardless of any theory. However, in order to derive information from observations it is necessary to identify what is being observed, which requires the presence of a relevant body of information.

Inductivism makes science look like it is no longer about things in the usual sense but rather about observations and their relationships. Can that be true? The fact of the matter is that trained scientists "see" much more in their observations than laypeople—they see "things" that other people might easily miss. As the philosopher of science Norwood Hanson puts it, "there is more to seeing than meets the eye." The same holds for the other senses: when a chemist, for instance, "smells" sulphur dioxide, a layperson "smells" rotten eggs. Although they have the same sense data, chemists know there are gases, and they know more about gases than laypeople, so chemists can learn more from their sense data. That's why we often look but not see, or listen but not hear.

(2) What comes with this is that inductivism portrays science as a nicely and steadily growing accumulation of obser-

vation statements—often eagerly called "facts" (§8). However, observations are very limited on their own. If observation statements are the foundation of science, then science can only talk about what we can observe—color, temperature, weight, movement, and so on—but never about their causes, which are often not observable, or about many other aspects. The history of science, on the other hand, shows us that research does not just keep adding more and more new observation statements to its present collection. Often, previous scientific accomplishments are being overturned. At times, theories that have served for a while are suddenly being abandoned. Sometimes, observation statements that seem to refute what was accepted before are set aside. In other words, the idea of a steadily growing fund of observation statements in science is hard to corroborate.

Nevertheless, the inductivist approach of science portrays science as an accumulation of singular observations. What laypeople call "things" are merely taken as bundles of observations. But the problem of this view is that science is not an encyclopedia of unrelated information—science is more than a mindless collection of observations. True, the job of scientists, as we all know, is to discover facts; but a haphazard collection of facts cannot be said to constitute a science—it's more like a broken down bookcase. Isolated particular facts may be known by direct observation, but scientists seek more than a mere record of such phenomena and facts—they try to understand them by formulating general laws which create a systematic relationship between them. They want and need to go beyond their observations.

In physics, for instance, they often speak of a "Grand Unified Theory" (GUT)—a theory that unifies into one theory the three non-gravitational forces, namely the electromagnetic force, the strong nuclear force, and the weak force. The aim of science is not merely collecting more and more interesting observations; science also wants to integrate them into a larger framework. Science is in search of a unifying picture.

(3) Inductivists assume that observation statements are the building blocks of science. But this raises immediately the question: which kind of observation statements? Inductivism assumes that all observation statements are basically equal— each one of them equally able to propel science. Just any? Of course not! Some are better than others. What is it that determines which ones are better? Perhaps the shortest answer is: that depends on the problem under consideration. If so, the problem under consideration comes *before* scientists make any observations. This explains why "good" scientists are considered experts in defining the formulation of their problem. "Divide and conquer" is a basic scientific strategy for attacking any difficult problem in science: take a large problem and divide it into smaller problems, then conquer the smaller problems one by one.

Usually scientists do this by using what they call *models*. They can even use different models to describe the same underlying reality. As the physicist Sean Carroll puts it, "Sometimes you might want to talk about a box of gas as a fluid with pressure and velocity, other times you might want to talk about it in terms of atoms and molecules." Models focus on what is considered relevant by omitting what is irrelevant. They come with ideal, abstract conditions which help scientists to study the behavior of any system under investigation.

To understand the motion of a billiard ball, for instance, it makes sense to treat the ball as if it were a perfect sphere of infinite rigidity, without worrying about its chemical composition, scratches on its surface, and all sorts of other details. This kind of idealization does not alter the fact that perfect balls, ideal gases, perfect conductors, ideal neurons, typical kidneys, or ideal populations do not really exist—they only do in the model of their respective fields. Beginning students of physics, for instance, quickly become acquainted with idealizations such as the notion of a frictionless surface, and with the fact that laws such as Newton's law of gravitation mathematically describe the behavior of bodies only in the situation where no interfering forces are acting on them—a situation that never actually holds, not even in the lab.

Because of this, models provide only representations of *aspects* of reality—perhaps rather relevant, but always partial. As the British philosopher of science John Dupré puts it, "The abstractions that work well in one context may eliminate what is essential in another." This way, models provide partial insight into the complexities of the real world. It is a truism that a "super-model" honoring all the details of our complex world would be as complex as the world itself. Models never fully replicate what they represent—that's what makes them models.

(4) Inductivism may give us the impression that all observation statements are equally *relevant*. It does not explain what the litmus color has to do with acids, rather than with carbohydrates. It does not explain what the expansion of iron has to do with temperature, rather than with atmospheric pressure. It does not explain why valves in veins went unnoticed as valves until the idea of blood circulation had been established. The inductivist typically does not answer these questions, does not even think to ask these questions. Yet, such questions need to be answered.

In other words, there are relevant factors and irrelevant factors. Induction has nothing to say about this, yet scientists don't bother with irrelevant factors or variables—that would be a waste of their time. But how do they determine which factors are relevant and which are irrelevant? Unfortunately, factors and circumstances do not come with neat little tags marked "relevant" and "irrelevant." So, what is to count as a relevant variation in the circumstances for boiling water, for example? Does variation in pressure matter? Yes, it does. Does the purity of water matter? Yes, it does. Does the method of heating make a difference? No, it doesn't. Does the time of the day make a difference? No, it doesn't. And we could add an unlimited number of additional factors. But who or what determines all of that?

What is relevant in scientific research has to be identified *before* any methods can be used. Since we cannot take *all* factors or all circumstances into account, we must have a *hypothesis* stating that the factors and circumstances we do take into account are the only relevant ones. Every scientific investigation begins with some kind of hypothesis, which determines which factors are considered relevant. In other words, there has to be something like a clear, central formulation of a problem. This delineates what scientists are interested in. One cannot observe every possible aspect closely, as inductivists might suggest. Therefore one must discriminate and try to select what is relevant, based on a hypothesis. Again, a hypothesis is needed before we can even speak of observation statements. There are so many factors involved that we can't just stare at them and probe each of them mechanically. Inductivism may have rules how to test them, but it offers no rules how to select which ones to test as being relevant.

(5) Inductivism starts with observation statements, but that may be a non-starter. Since scientific research is most of the time a mine-field of "hidden" variables, scientists need to control their variables carefully before making observation statements. Unlike what inductivists seem to suggest, scientists don't just hunt for more and more observations, under various different settings, but they limit themselves to a strictly *controlled* setting. They usually do so by removing the process or

object under investigation from its natural context, usually in a lab.

Expressed more technically, during the experiment, interfering variables have to be ruled out or controlled. Therefore, scientists limit themselves to a simple setting. To put this in an image, an experiment needs to take place in the "test-tube-like shelter" of a laboratory, removed from the complexity of nature. In a test tube (*in vitro*), for instance, a biological experiment is easier to keep under control than inside the body (*in vivo*). And something similar holds for the other sciences. Physics, for instance, often deals with "closed systems," which are basically systems "in a test-tube."

The need for factors that are relevant to what scientists are studying is usually done through the model they use. Again, the model helps them to keep strict control over factors that are not considered relevant, yet might interfere with their search. In other words, scientists create a simplified model of what they are studying. Reducing complex entities to a manageable model related to an analyzable problem is a successful way of doing research. This means there is always a "ceteris paribus" clause involved to the effect of "everything else being constant." The consequence of this clause is that other (possibly) interfering variables have to be kept under control or are assumed to be constant. Even the "ceteris paribus" clause is only understandable and possible under the assumption that not everything is equally relevant. No observation statement can start this process on its own.

(6) Inductivism basically leads to establishing correlations, which are generalizing statements evolving from "Some X's are some Y's" to "Many X's are many Y's." However, correlation should not be confused with causation. The point is often summed up in the maxim, "Correlation is not causation." Many people heard this maxim already in high school. Just because

two factors are correlated does not necessarily mean that one causes the other. Correlation may be useful for prediction, but it does not always give us an explanation of a causal connection, if there is any. Besides, correlation is symmetrical, but causation only goes one way.

In general, it is much easier to establish correlation than to prove causation. A case in point is the use of a brain scanning technique called functional magnetic resonance imaging (fMRI)—an MRI procedure that measures brain activity by detecting associated changes in blood flow. Neuroscientists have discovered that certain brain regions light up on an fMRI during specific mental activities. However, what we have here is a matter of mere correlation. Based on this, we do not know whether changes in fMRI cause those mental phenomena, or whether mental activities cause changes in fMRI, or whether mental and neural phenomena have a common underlying cause, or whether their correlation is merely a case of accidental associations.

A more familiar case is the correlation between smoking tobacco and the occurrence of lung cancer. In order to call tobacco a mutagen or carcinogen, we need to prove causation, not just correlation. That's easier said than done. It could well be that the increased use of tobacco is as much associated with lung cancer as is the increased use of nylon stockings, for instance. In addition, there could be many other causal factors involved: the increased rate of lung cancer could have been the result of better diagnosis, more industrial pollution, or more polluting car exhaust on the roads; perhaps people who were more genetically predisposed to smoking were also more susceptible to getting cancer. It took a large study involving more than 40,000 doctors in the UK to show conclusively that smoking really does cause cancer.

But even if we can prove that smoking causes lung cancer,

that does not mean that everyone who smokes will develop it, for smoking is not a sufficient cause of lung cancer. Nor is smoking a necessary cause; people who do not smoke can still develop lung cancer because there may be other causal factors involved. In other words, inductivism is too simple a strategy to deal with such issues.

(7) Another assumption of inductivists is that scientists can just follow simple rules to identify causal relationships between variables. But that's too simple a statement. Elementary cause-and-effect connections, such as between fire and pain, or rainfall and harvest, can be made by almost everyone, but science deals with much more complicated cause-and-effect relationships—such as the common cause behind falling apples and orbiting planets. Mills' rules may help us to *demonstrate* causal connections, but they do not help us to *discover* causal connections. His rules work only if we know already all the factors that might cause the outcome, so we can eliminate some of them as irrelevant.

The following example may clarify this in a simplified way. Imagine, you want to find out whether the headache someone developed is caused by "gin on the rocks," or by "whisky on the rocks," or perhaps by "rum on the rocks." According to Mill's rules, you should extensively experiment with these three drinks... to come to the surprising conclusion that your headache is caused by ice cubes, because that is what all these drinks have in common. This conclusion seems quite reasonable, until you come to know of the more-embracing concept of "alcohol"—being a generic term for gin, whisky, and rum combined. Mill's rules do not help us to come up with that concept.

To arrive at a new concept, a mental leap is needed. Not any kind of inductive rule can achieve this for us. Searching in science is more based on creativity and imagination than on observation and manipulation. In the search phase of scientific research there is more need of provisional ideas than of logical tools. This explains why no one saw the similarity between falling apples and orbiting planets before Isaac Newton had made the connection—and it took him a while too in order to come up with the concept of gravity.

(8) Inductivists also tell us that the number of observations determines how secure our scientific knowledge is. This inevitably raises new questions. How often, for instance, do we need to repeat the observation that water boils at 100°C? Ten times, a hundred times? It is hard, if not impossible, to tell. Obviously, to secure *all* possible observations, we would have to await the end of the world, so to speak. Besides, the effect of adding another new favorable observation will generally become smaller as the number of instances grows.

On the other hand, having observed the sun to set each day on many occasions does not guarantee us that the sun will set every day—the sun never sets in the Arctic or Antarctic, for instance. Even a probabilistic approach won't work—something like, "There is a 99% chance for the sun to set." The probability of a finite number of observations divided by an infinite number of possible situations remains zero, no matter how large the number of observations is.

In general it's true that the greater the number of instances observed, the lower the probability is of finding an exception. But so long as there are unobserved instances, there is always the possibility that the inductive conclusion will be shown to be false on further investigation. The philosopher Bertrand Russell made fun of the inductivist method by introducing his "inductivist turkey." On his first day at the turkey farm, this turkey discovered he was fed at 9 AM. However, being a good inductivist, he did not jump to conclusions, but waited for a larger number of similar observations. Each day the turkey added another observation statement to his list, so he came to the conclusion that all days would be like that. The turkey was right until Christmas time came.

Even if someone objects that the method of induction has been rather successful so far, we should mention David Hume's objection arguing that there is no way we can justify induction by showing it has worked successfully in the past. To justify induction this way is a form of circular reasoning because it employs the very kind of inductive argument that we are trying to justify. We cannot use induction to justify induction. Perhaps we may conclude from this that the number of repeated observations is not that essential for scientific research at all.

(9) Inductivists want us to believe that observations are the clear-cut, unambiguous, elementary building blocks of science. But the problem for inductivism is that observations are not what they appear to be—they are not clear-cut or neutral or assumption-free. Observations are not solely determined by what observers see on the retina of their eyes. What scientists see through microscopes or telescopes, for instance, is very different from what laypeople discern—yet they may have the same images on their retinas. It's the same story for what X-ray pictures reveal; medically trained viewers see things differently, and arguably better, than beginners. Even if we assume that observations are the same for everyone—which they are not, except for images on their retina—we still need to acknowledge that observations statements involve theories of various degrees of generality and sophistication. In short, all observation statements must be made in the language of some theory, however vague or primitive.

In contrast, the simplistic view of science has it that most experiments are based mainly on analog and digital readings produced by measuring equipment—observations, that is. However, this is a rather deceptive view. Before a "signal" can be read off, it has to be separated from the "noise" that muffles it. This tedious process is done by spotting each interfering factor, step by step. As said before, the setting up of an experiment has to be such that interference has been completely precluded. The decision that there is no further interference left and that a certain reading is a reliable measurement is always a provisional one, for scientists must necessarily rely on their scientific expertise, their experience with measuring equipment, their theoretical background, as well as the judgement of their colleagues in the scientific community.

For a long time, for example, it was thought that the purity of elements was only determined by chemical techniques. The English chemist William Prout had observed in 1816 that the atomic weights measured for the elements known at that time appeared to be whole multiples of the atomic weight of hydrogen. However, in particular the atomic weight of chlorine, which is 35.45 times that of hydrogen, could not at the time be explained in terms of Prout's hypothesis which predicted a weight of 35. Then, by 1925, the problematic chlorine was found to be composed of the isotopes Cl-35 and Cl-37, in proportions such that the average weight of natural chlorine was about 35.45 times that of hydrogen. Prout's hypothesis turned out to be correct for atomic masses of individual isotopes, but "chemically pure" elements may not be "physically pure." Apparently, purifying techniques can have quite an impact on test implications and predictions. From now on, a pure element contains atoms of only one isotope, so every part of the substance has the same physical properties. Neutral observations hardly exist.

The problem of observations is not only found in observation techniques. The observations themselves are rather ambiguous besides. By using the word "empirical," one may create the impression that scientific research is based on a simple recipe: "Just open your eyes and ears!" If that were true, research would be a simple process of just picking up regularities, connecting relations, and general laws waiting to be unveiled. However, our eyes do not function like a camera, nor our ears like a microphone. The greatest scientist, as we all know, is not the one just recording everything on tape or on film. When Charles Darwin wrote to Henry Fawcett about geologists who felt obliged to observe and not to theorize, he was right in saying, "A man might as well go into a gravel pit and count the pebbles and describe the colors. How odd it is that anyone should not see that all observations must be for or against some view if it is to be of any service."

Even if we were to regard our eyes as producing "pictures" in the way a camera does, we would not have solved the problem that those pictures in turn have to be viewed and interpreted next. "Seeing" is a matter of structuring a collection of "dots and lines"; and "hearing" is a matter of structuring a series of sounds. The truth is not in nature waiting to reveal itself, but we depend on an imaginative preconception of what the truth might be. As an old saying goes, "We are prone to see what lies *behind* our eyes rather than what appears *before* them."

In other words, reasoning from observation to observation assumes that observations are clear-cut, preset elements underlying scientific knowledge and explanation—which is an assumption that has to be seriously questioned. The physicist and philosopher of science Karl Popper used to say that the command "Observe!" does not make any sense, since no one would know *what* to observe. His point is that scientific theories just do not and cannot spontaneously emerge from observation. We do not "have" observations—like we have sensorial experiences—but we "make" observations. Philosophical giants such as Aristotle and St. Thomas Aquinas would put it this way: all we know about the world comes through our senses, but this is then processed by the intellect that extracts from sensory experiences that which is *intelligible*. That is the reason why a camera cannot make observations. Cameras record, but they do not observe anything. A surveillance camera, for instance, automatically "observes" every single detail because it does not know what to observe. That is why cameras cannot replace scientists—they may help them but cannot replace them.

The problem with any kind of images or pictures is that they do not show us observations until we give some *interpretation* to the things and events we see on the picture. This requires concepts (§11). It is through concepts that perceptions become observations. It is through concepts that events become part of a certain category beyond a particular event—which makes for an observation. It is through mental concepts that we transform "things" of the world into "objects" of knowledge. Concepts change perceptions and experiences into observations, thus enabling humans to see with their "mental eyes" what no physical eyes could ever see before. To be sure, all we know about the world does come through our physical senses, but this is then processed by the immaterial intellect that extracts from sensory experiences that which is intelligible in conceptual terms. Without concepts we listen but do not hear, we look but do not see, we are "blind" with our eyes open. A concept is a kind of preconception of what some part of the world looks like.

(10) Another assumption of inductivism is that we need a series of *similar* observations leading us to generalizing induction. However, the act of generalizing is based on the assumption that objects and events are in fact similar and look alike in certain respects—e.g. "being magnetic," "being acidic,"

"being cellular," "being infectious," "being toxic," and so on.

The problem is, though, that this similarity is not visible until we know already what it is that "similar" cases have in common and what makes them similar—which requires the proper concepts. We need to identify first what is relevant to our problem, because similarity cannot be established until it has been identified in a word, or actually in a concept. Scientists cannot mechanically infer from a few cases to all similar cases until their similarity has been conceptualized first. Since things can be "alike" in numerous ways, we need a unifying concept of similarity first before we can classify and categorize things as similar. Before we can "notice" a carnivore, we need the "notion" of a carnivore to begin with. Without the concept of "carnivore" we cannot see carnivores.

Take this simple example. No matter how closely one studies colliding billiard-balls, orbiting planets, or weights on springs, one will never arrive at the concept of "mass" through observation alone. Neither is it possible to teach others the meaning of "mass" merely by pointing to such phenomena. Michael Augros tells us how his physics teacher in high school explained the difference between *mass* and *weight* with the question, "Would you rather lift my car or push it?" That's how he learned that mass resists your push, and weight resists your lift. This teacher actually made an effort to explain these concepts without pointing at something. That step in understanding seems to be necessary before the equations of physics can even begin.

Apparently, the meaning of concepts is not acquired through observation. When observers see several red things, they are supposed to discern that their common element is redness. However, a set of red things does not select itself. The criterion which includes certain perceptual experiences in the set of red things presupposes the very concept, redness, which it is meant to explain. Mere observations of red objects do not generate the concept of redness. Pointing at a set of red things does not automatically generate the concept of redness. The truth of the matter is that observations do not create their own observation statements.

More in general, "similarity" in science stands or falls with concepts such as "gravity," "magnetism," "cell", "infection," "toxicity," and "gene." Without cognition there is no re-cognition. Hence, there is always a conceptual leap involved. Take the case of a sickness caused by bacterial infections. For the sickness to be explained that way, somebody has to come up with the idea or hypothesis of a bacterial infection. Before Robert Koch and Louis Pasteur had finished and published their experiments, no one would have ever thought human sickness could be caused by an infection with bacteria. Actually, the problem is that infinitely many factors may qualify as the potential cause of a certain phenomenon. Mill's rules of induction only work when we have before us *all* and *only* the facts relevant to the solution of our problem. But that is quite an assumption; most of the time we do not! Only someone who has a number of possible causes in mind—which implies a hypothesis based on concepts—can use inductive rules to eliminate erroneously assumed causes.

Imagine that the concept "cell" were really discovered by perceiving what a set of cellular objects has in common. This would suggest that observers are able to discern, in their observations under the microscope, a common element of a set of things and come to understand this common element as "being a cell." However, in order to decide which observations are included in the set and which are excluded, we need a criterion which says that only observations of "cellular" objects are to be included in the set. Again, this account presupposes the very concept, cellularity, the acquisition of which it is meant to explain. Things cannot be "alike" until the similarity referred to has been established beforehand. That is the reason why we can never be certain that showing someone a collection of cellular objects will make that person identify the set as a collection of cells. Biology teachers can testify to this.

Anything can be pointed out, once it has been recognized; but not everything pointed out is going to be recognized immediately. Thus, we end up with some kind of paradox. On the one hand, there is no searching unless there is some knowledge; on the other hand, complete knowledge would make searching redundant. It is this seeming paradox that makes us realize that gaining knowledge is a never ending alternation of searching and finding. Knowledge is initially provisional, fragmentary, and partial, although it is in the process of becoming more and more comprehensive. Finding is often a consequence of goal-oriented searching—and in this respect it resembles the work of paleontologists and archeologists who start digging where they expect to find something. Science is a matter of methodological searching. How could we search if we had no idea of what to search for?

The problem we have here is probably best expressed by the ancient philosopher Plato, although in a slightly different context, when he said: "How would you search for what is unknown to you?" Plato noticed a seeming paradox here: we are in search of something "unknown"—otherwise we would not need to search anymore—and yet it must be "known" at the same time—otherwise we would not know what to search for, or would not even know if we had found what we were searching for. This is the reason why we need concepts and hypotheses in science, as those can open our eyes for similarities we would not have been able to see without them. Concepts act like search-lights—they help us search, and perhaps they make us find.

Seen in this light, concepts are miniature theories—theories in the making. Most concepts have an intricate web of connections with other concepts—for instance, in case of a circle, with concepts such as "radius" and "diameter." As a result, concepts go far beyond what the senses provide—they transform "things" of the world into "objects" of knowledge, thus enabling us to see with our "mental eyes" what no physical eyes could ever see before. From this follows that observation is seen as a "*concept*-laden" or "theory-laden" phenomenon that does not originate from "barren observations at the bottom" but rather from "reason and intellect at the top."

Yet, you might counter, sometimes science seems to start from haphazard observations. Isn't it true that mere "luck" does play a role in research? Indeed, sometimes a scientist may find something without really searching. Frequently, when relating some new finding, you may hear scientists say almost apologetically, "I came across it by accident." Most of the time we call this "a stroke of luck." Wouldn't that make for an observation without any expectation?

Not really. We have to realize, as Louis Pasteur used to say, that luck is only productive for "prepared minds." It is like with paleontologists and archeologists; they don't just start digging where they happen to be; instead they have some idea of where they have a good chance to find what they are looking for. As the motto over the entrance to Harvard Medical School has it, "Chance favors only the prepared mind." Think of the periodic table in chemistry. Dmitri Mendeleev had to conjecture that atomic weight was the attribute according to which the elements should be ordered. Interestingly enough, he made his discovery just a few years after the concept of atomic weight had been clarified. That's what prepared him somehow for his periodic table.

After the ten arguments mentioned above against the

simplistic view of inductivists, it's time now to wrap up our discussion. Observations cannot generate their own observation statements, for those statements require concepts that observation cannot provide. We had to come to the conclusion that it's hard to believe that science works and advances along inductivist lines. It is highly unlikely that there are simple "rules of induction" by which concepts, hypotheses, and theories can be mechanically generated from empirical data— let alone from so-called observations. Logic provides us with formal means that allow us to check an argument after it has been made. The rules of logic are not rules of discovery but at best rules of validation. Logic does not allow us to detect statements before the event, but only to check them after the event and thus to provide them with a seal of approval, if possible. In other words, inductivism, empiricism, and positivism give us the wrong impression of what science is like. There has got to be a better view of what science really is and does.

3. A More Realistic View

As we discovered in the previous chapter, scientific research is first of all a matter of asking the right questions—by means of new concepts, models, hypotheses, and theories. There is no such thing as seeing-in-a-neutral-way, or observing-without-expectation, so we found out. All observation is somehow "concept-laden" and "theory-laden." How could a "meaning-less" observation ever be relevant to testing a scientific theory? The best way to search is to have an idea of what you are looking for—the idea may be wrong, but it's very hard, if not impossible, to just search "blindly." The search phase thrives on ideas; without concepts and theories in the search phase, science would indeed be blind.

Seen this way, scientific research may be described as a "dialogue" between scientists and nature—that is, a dialogue between *possible* observations (hypotheses) and *actual* observations (facts). This process is done in what has been called the *empirical cycle*—which is admittedly a textbook simplification. According to the empirical cycle, research often starts with certain observations that are seen as problematic in the light of some theory. To explain these problematic observations, scientists come up with a hypothesis to find an explanation for them. That seems to be an important starting point in science. But let's make clear first that a hypothesis is more like an "invention," which is very different from a "discovery." To find out whether the invention qualifies as a discovery, we need to derive test implications from this hypothesis—so as to make sure the hypothesis is more than an invention.

Science stands or falls by deriving test implications from its hypotheses, which are basically predictions that can be tested either in the field (empirical testing) or in the lab (experimental testing). They are of the type "If the hypothesis is true, then..." or "If the hypothesis is not true, then..." Simply put, the outcome of each test implication can go one of two ways: the test decides either for or against the hypothesis; in other words, the hypothesis either makes true predictions or false predictions, which either confirm or refute the hypothesis under investigation.

An example may help explain this. Murray Barr had discovered that one of the two X-chromosomes in female cells is in a condensed form. Mary Lyon came up with the hypothesis that one of the two X-chromosomes will be randomly inactivated or silenced into a "barr body," thus preventing females from having twice as many X-chromosome gene pro-ducts as males, who only possess a single copy of the X chromosome. Based on this hypothesis, she could predict several new observations. The randomness of the inactivation was confirmed by the coloration of tortoiseshell female cats which have a variegated fur pattern. The black and orange alleles of a fur coloration gene reside on the X-chromosome. Cells in which the chromosome carrying the orange allele (O) is inactivated express the alternative non-orange allele (o), whereas cells in which the non-orange allele (o) is inactivated express the orange allele (O). This not only creates a "genetic mosaic," but also a fur color pattern in females.

The Lyon hypothesis had many more test implications. One of its predictions was that heterozygous females with both a functional and a dysfunctional X-linked allele for certain enzymes would only produce half of the amount of enzymes compared to females with both functional alleles. Another prediction was that individuals with an abnormal number of X-chromosomes—for example, men with Klinefelter syndrome (XXY) or women with a third X-chromosome (XXX)—would produce the same amount of gene products as a normal male, unless they happen to have a dysfunctional allele on the activated X-chromosome. These and other test implications were confirmed and thus strengthened the Lyon hypothesis.

Seen this way, scientific research is like a dialogue between the possible and the actual, between fiction and reality, between invention and discovery. If the test implications come true, we receive more *confirmation*; but if a test implication is not confirmed, we speak of *falsification*. The empirical cycle is like a "game" of questioning and answering, of searching and testing, of trial and error, by falling over and picking oneself up again. It is like a dialogue between subject and object: the subject (the scientist) asks a question couched in a hypothesis, and the object (nature) gives an "answer" phrased in terms of test results. It is a process that could be likened to the investigation of a detective. Seen this way, research may be seen as a "simple" process. But a few remarks need to be made about this process.

First, most research starts with a hypothesis—rather than with the observation statements of inductivism. But we need to be aware of what this entails. In science, discoveries typically start as inventions—usually called hypotheses. But it needs to be stressed that not all inventions lead to discoveries. To use an analogy, the person who invented "Atlantis" did not discover Atlantis; it remains a legendary island until further notice. The same in science: most inventions do not make it to the stage of discoveries. The invention of Vulcan as a planet between the Sun and Mercury never made it to a discovery.

Nonetheless, some scientists think they have made a discovery when all they have in mind is an invention, a hypothesis. However, a hypothesis is only an invention in the scientist's mind until it has been shown to be also a discovery in reality. The late Nobel Laureate and physiologist Peter Medawar gave a wise advice to a (young) scientist: "the intensity of the conviction that a hypothesis is true has no bearing on whether it is true or not." Because of this, science has an important rule for hypotheses if they are to be of any use: they must have observational consequences for them to qualify as scientific.

In other words, scientific hypotheses are basically "happy guesses" or "bold conjectures," which are not derived or discovered from observation statements, but they are invented first by some scientist. Obviously, they are not "wild guesses"— that's why a complete novice will hardly make any scientific discovery. Yet, "guesses" guide us as to where and what to search. Even in simple cases such as "Iron expands with heating," we have a "hidden" hypothesis stating that a certain variable, the length of the iron rod, is a function of one single other variable, temperature. It is through deduction that we derive from this hypothesis predictions that can then be tested in the field or in the lab as to whether they are inventions that do lead to discoveries. Research is first of all a matter of asking the right questions—by means of new concepts, models, hypotheses, and theories. But they need to be tested further in the field or in the lab.

Second, to explain certain scientific puzzles, several hypotheses may qualify simultaneously, for the simple reason that the same phenomena can be explained by different hypotheses. Think of a few observation points marked in a graphic and try to draw a curve through these points. You will find out there are many, even infinitely many, possibilities, unless you have a preference for the smoothest curves only. Or take this simplified case: if all I know is that someone spent $10 on apples and oranges and that apples cost $1 while oranges cost $2, then I know that this person did not buy six oranges, but I do not know whether he or she bought one orange and eight apples, two oranges and six apples, and so on. In other words, there might well be other theories that are also well confirmed by the very same body of evidence.

Competing hypotheses are rather common in science. To use some simple, historical example, the water in a water tower is "sucked up" by a "horror vacui" (nature's abhorrence of a vacuum) according to Galileo's hypothesis, but it is "pushed up" by atmospheric pressure according to Torricelli's. Either explanation is theoretically possible, until we come up with test implications that may favor one of the two hypotheses over the other.

In fact, Blaise Pascal came up with an ingenious test implication to decide between the two hypotheses. He reasoned that if the height of the mercury column in Toricelli's barometer depends on the pressure of the air, then its length should decrease with increasing altitude. So he had his brother-in-law carry the barometer to the top of a nearby mountain and found his hypothesis confirmed. Was that a decisive proof against the concept of a "horror vacui"? Not necessarily so, for perhaps such abhorrence decreases with increasing altitude too. But that would rather be considered an *ad-hoc* adjustment—an escape strategy—to save a theory from being falsified.

A more complicated example can be found in the early stages of astronomy, where at least three theoretical frame-works were able to explain the phenomena seen in the skies at night: the geo-centric model of Ptolemy with the earth in the center, the helio-centric model of Copernicus with the sun in the center, and the model of Tycho Brahe who had the Moon and the Sun revolving around the Earth, but the other planets (Mercury, Venus, Mars, Jupiter, and Saturn) revolve around the Sun; so he had the Sun with those planets together revolve around the Earth.

Interestingly enough, the model of Ptolemy and the model of Copernicus were in fact mathematically identical for all observations available at the time. Besides, Tycho's system also fit all the prevailing data, including Galileo's discovery of the phases of Venus which revealed that the Sun, the Earth, and Venus were sometimes configured in ways not possible in the Ptolemaic system. Eventually, Tycho's system would only fall out of favor on theoretical grounds, almost a century later, when Newton's theory of gravity (1687) made it inconceivable for the solar system to orbit the small mass of the Earth.

Third, science has an important rule for hypotheses if they are to be of any use: they must have observational consequences for them to qualify as scientific. If the test implication comes true, we receive more confirmation; but if the test implication is not confirmed, we speak of falsification. However, when test implications turn out to be true—that is, if the hypothesis makes true predictions—this does not mean the hypothesis has been *proven* to be true too, at least not in the strict sense of verification. The best we can say is that the hypothesis has become more likely to be true. We don't get verification but only confirmation. Verification is in essence an unattainable goal. In science there is just no place for "knowing for sure." The most scientists can claim is that the hypothesis is extremely likely after they have received many similar confirming test results and predictions.

So this brings us basically back to the same problems we encountered with inductivism. Hypotheses hold in essence for an unlimited, infinite number of similar cases, but there is just no logically guaranteed way of reasoning from *some* singular instances to *all* similar cases, as we found out. Induction does not have the logical power of deduction. No matter how many different confirming test results we have, there is no guarantee the hypothesis is absolutely true. Yet, when more and more test results turn out positive, the hypothesis is on its way to becoming more and more convincing. This made Bertrand Russell describe science as follows, "its method is one which is logically incapable of arriving at a complete and final demonstration."

Fourth, because of the limited inductive power of confirmation, it has been claimed that scientists should exclusively go for the deductive power of falsification. Falsification is basically a logically safe, deductive way of reasoning: if we find one black swan, a previous inductive conclusion stating that all swans are white has been deductively and conclusively falsified. Or more specifically, the reasoning is as follows: if the hypothesis is true then its test implication is true; well, it turns out its test implication is *not* true; therefore the hypothesis *must* be false. In other words, as a consequence, the initial hypothesis has to be dropped, or at least revised.

No wonder, this idea has made many scientists favor falsification over confirmation, making them demand that a good scientific hypothesis must be falsifiable. In other words, we may not be able to prove, but at least we can disprove, so they say. The main proponent of this idea was the legendary Karl Popper, who put it this way: "Every 'good' scientific theory [...] forbids certain things to happen." Albert Einstein said something similar: "No amount of experimentation can ever prove me right; a single experiment can prove me wrong." Hence, scientists should always be ready to take "no" for an answer when falsifying evidence points that way. They realize that no amount of data can really prove a theory, but even a single key data point can potentially disprove it.

Think of the following example. In the 1930s, new experiments threatened one of the core principles of physics, known as the conservation of energy. The data showed that in a certain kind of radioactive decay, electrons could fly out of an atomic nucleus—even though the total amount of energy in the reaction should have been the same each time. That meant energy sometimes went missing from these reactions, and it wasn't clear what was happening to it. The physicist Niels Bohr was willing to give up the principle of energy conservation. But the physicist Wolfgang Pauli wasn't ready to throw in the towel. Instead, he came up with an outlandish particle. This new particle could account for the loss of energy, despite having almost no mass and no electric charge—so it was basically invisible at the time. Nonetheless, rather than agreeing with Bohr that energy conservation had been falsified, the physics community embraced Pauli's hypothetical particle—which came to be known as a "neutrino."

Here is another example. Einstein's general theory of relativity was a bold conjecture in 1915 because at the time background knowledge included the assumption that light travels in straight lines. It was Einstein's new hypothesis that gravity bends the path of light. If it can be demonstrated that a ray of light passing close to the Sun is deflected in a curved path, then it is not true that light necessarily travels in straight lines. Einstein took the risk that his hypothesis could be falsified, although in fact it was not. What to do, however, if it would have been falsified? Would that have been the end of his theory of relativity?

Not necessarily so, for there are some hidden problems with falsification. When Einstein said, "a single experiment can prove me wrong," he was certainly right if the emphasis is on the word "can"—but not automatically so. It is always possible to reject falsification for the simple reason that falsification is based on an *observation* that is counter to what the hypothesis or theory had predicted. But as we found out in discussing inductivism, there are no "clean" or "pure" observations—they are most of the time "concept-laden" and "theory-laden."

So ultimately, even falsifying evidence depends on verification. But if final verification is beyond our reach, so must falsification be. Therefore, it is always possible to reject the counterevidence as spurious. The main point is that counterevidence is based on an observation statement, which means that counterevidence doesn't automatically qualify as falsifying evidence. It is always possible to question the validity of any counterevidence, since evidence is based on an observation statement that only qualifies as evidence when it is beyond any doubt.

An added problem of falsification is that it is very unusual in science that a single theory is being put to the test. Almost every hypothesis or theory in science is tied to a certain model, which means that the hypothesis is only applicable under the conditions and presuppositions as laid down in the model. Apart from these boundary conditions, the hypothesis is a castle in the air. Add to this that in most sciences observation relies heavily on complicated instruments, which makes the impact of theories even more dominant. Most falsifying observations owe their existence to some kind of measuring equipment and some theoretical background—and either one of these may need some adjustment. Only the experimenter knows the many little things that could have gone wrong in the experiment.

This last part of the falsification problem has led to what has become known as the Duhem-Quine thesis, which states that hypotheses are not tested in isolation but always as part of whole bodies of interconnected theories. So there might be several reasons why we may encounter falsification, given the fact that there are many other assumptions involved. Hence, falsification only tells us that "something" is wrong—which may be the hypothesis itself, or the assumed boundary conditions, or the assumed theoretical and/or instrumental context of the counter-evidence. Therefore, a case of falsification must lead to a threefold conclusion: the hypothesis is wrong, or an interfering factor is involved, or the counter-evidence is not a fact (§8).

In other words, falsification does not make a falsifiable hypothesis or theory automatically falsified on the first hit. We may have thought we had a new motto for science—"Although science cannot prove, it can disprove"—but now we have to add that science cannot even "disprove" in a rigorous way. When nature says No to our tests, it is not exactly clear what exactly it says No to.

This outcome may seem to leave us hanging in the air. Is there any reason why scientists do or do not accept falsification? Many answers have emerged. One explanation was given by Thomas Kuhn (1922-1996). He introduced the term *paradigm*, which has been defined in various ways, but it typically stands for a collection of rules on how to solve scientific puzzles. Aspiring scientists are being brought up with a particular paradigm, making them feel attached to the paradigm they were brought up with. The reason is that they acquired knowledge of a paradigm through their scientific training—that is how they learn their standards, by solving "standard" problems, performing "standard" experiments, drawing "good" conclusion, and eventually by doing research under a supervisor already skilled within the paradigm.

Because of their training, scientists are usually unable to articulate the precise nature of the paradigm in which they work, until a need arises to become aware of the general laws, simplified models, metaphysical assumptions, and methodological principles involved in their paradigm. This awareness may force them to think "outside the box." It is also part of a paradigm to determine which falsifications are considered "real" and which are seen as "spurious."

Another, arguably better, explanation of what to do in case of falsification was given by Imre Lakatos (1922-1974). His key word is *research program*, which is a strategy of programmed refutability standing midway between dogmatism and anti-dogmatism. Every research program has a provisional "hard core" which is protected against falsification; it tells scientists what paths of research to *avoid* and is called the "negative heuristic" of the program. Apart from this, there is a "protective belt" of auxiliary hypotheses which have to bear the brunt of systematic tests and may have to be adjusted; this belt is called the "positive heuristic" and tells scientists what paths of research to *pursue*.

The two explanations given by Kuhn and Lakatos are rather theoretical, but they can sometimes explain why scientists make certain decisions the way they do, especially so when it comes to falsification.

Here are some examples of how paradigms or research programs did change in the history of science: (1) the transition in cosmology from a Ptolemaic cosmology to a Copernican one; (2) the acceptance of Lavoisier's theory of chemical reactions and combustion in place of the phlogiston theory; (3) the transition from Newtonian gravity to the Einsteinian theory of general relativity; (4) the change from classical mechanics to the development of quantum mechanics; (5) and so many more.

What can we conclude after all of this? Science is a tentative enterprise, leading to acceptance, revision, or rejection of a hypothesis or theory, or even of an entire paradigm or research program. What is accepted today may be modified or even discarded tomorrow. Because the empirical cycle is a never-ending, cyclical process, science never ends. Don't confuse this empirical cycle with a vicious circle; it is more of a spiral than a circle—hopefully a cycle spiraling upwards and not downwards. Although scientific research is not inductive in the narrow sense (§1), it is still inductive in a wider sense—it accepts hypotheses on the basis of data providing inductive support, but never deductively conclusive evidence.

4. Ten Examples from the History of Science

What historians of science like to do is describing scientific methods and rules, instead of prescribing them. However, this is easier said than done, as rules and methods can only be described once it is known where they have been applied. At first sight it seems clear where they have been applied, namely in science! But this raises the following question: What is considered to be science, and what is not? Could it be that science is characterized by the very rules and methods we want to describe? If so, we are about to pre-scribe the rules before we are able to de-scribe them! It is not easy to escape this vicious circle. Besides, the scientific community may have changed some of its rules and standards for carrying out research. Therefore, a purely historical approach is not of much help either, for what used to be called science in the past may no longer be called science at present.

Let's leave this circular argument for what it is worth. In what follows, we will discuss and analyze ten examples from the history of science so as to see whether our discussion about science thus far has any footing in what scientists actually have been doing. The examples are such that they can easily be followed and understood by everyone. There are many other, more complicated cases, but in this book, I would like to stay away from those.

Case #1: Harvey – Blood Circulation

Let's start with the simple case of how William Harvey discovered the existence of a closed blood circulation. He did so at a time there was much confusion about the bloodstream. Strange conceptions had been around for centuries.

Galen of Pergamum (c. 150 A.D.), for instance, a Greek physician, surgeon, and philosopher in the Roman Empire, had a rather peculiar conception of the bloodstream. It was his conviction that our bloodstream is not based on a closed circulatory system, but instead he believed the expanding and contracting movements of the heart cause the blood to move up and down the blood vessels, as it is with the tides—some kind of two-way-traffic in the blood vessels. However, in order to allow blood to move from the veins to the arteries, Galen had to postulate pores in the septum of the heart.

Andrea Cesalpino (1509-1603), an Italian physician, was one of the first scientists to acknowledge the problems inherent in Galen's system. He tried to deal with the valves in the veins—which the anatomist Hieronymus Fabricius (1537-1619) had called "little mouths"—by suggesting that they regulate how much blood is going to seep from the arteries back into the veins. Somehow he was preparing the way for Harvey's closed system.

Michael Serveto (1511-1553), a Spanish physician, tackled another problem. He wondered how blood can flow from the right to the left part of the heart through pores that were never seen. It seemed more acceptable to Serveto that blood should flow from one part of the heart to the other by way of the lungs—and thus he introduced a one-way-traffic between heart and lungs. Ironically, in postulating a partially closed system, he replaced the unseen pores in the heart by unseen tiny vessels in the lungs.

It was William Harvey (1578-1657) who deserves the prize for finally reshuffling the pieces of the puzzle into one new system. An idea was born! Harvey's new idea was not really a discovery but more of an invention. He never saw the connecting blood capillaries needed for a closed circulation. We do not know what exactly gave him his revolutionary idea. Was it the thought of an outlet needed for a pumped inflow, or rather Aristotle's idea of perfect circular motion, or perhaps the developing technology of pumps? We don't really know.

Whatever it was that guided him, Harvey hypothesized two one-way circulations controlled by a heart pump. Apparently, the capillaries did not show the closed circulation of blood, but it was the other way around: the very theory of a closed blood circulation allowed him to deal with many anomalies of the older views. Only eventually would this theory make the vessels visible when better microscopes would become available later on. What we can learn from this is that Harvey's new theory certainly was not based on the observation of capillaries, as inductivism would like it.

By hypothesizing two one-way circulations ruled by a heart pump, Harvey could solve several anomalies: no unseen pores anymore, no tides anymore, no accumulation of pumped inflow anymore, and no useless valves anymore! At the same time, however, Harvey had introduced his own, new problems. How can arterial blood become venous blood, and where exactly does this happen? As Galen had created pores in the septum, Harvey needed a different kind of connections to obtain a closed system.

His contemporaries were well aware of Harvey's new problem. An external referee in the British Medical Research Council at the time had to write a comment on a grant application submitted by Harvey. The referee's report says: "He claims never to have observed pores or holes in the heart [..] By contrast, I have never seen any evidence for the existence of blood vessels providing a link between arteries and veins in the periphery of the body, as is stipulated by the scheme of Dr. Harvey; such vessels would have to be of an extremely fine calibre to escape the eye and would pose an insurmountable mechanical resistance to the flow of blood." That was indeed a real problem for Harvey's hypothesis.

Although Harvey never saw the connecting blood capillaries needed for a closed circulation system, he did not accept this as falsifying evidence for his new model. Instead Harvey held on to his new concept, his new hypothesis, and his new model. Only time would tell that his bold conjecture was indeed correct.

Case #2: Redi – Rotting

The centuries-old theory of spontaneous generation, dating back to Aristotle, holds that living organisms can spontaneously arise or develop from nonliving matter. Evidence for this was, for instance, the spontaneous appearance of maggots in meat, or the spontaneous development of mice in stored grain. According to this theory, living material can die, but "vital forces" cannot; these forces are even able to revive dead material that has undergone rotting, decay, or decomposition. How did science deal with this idea?

In the seventeenth century spontaneous generation was still a respectable scientific theory. One of the discoveries made by the German natural scientist Athanasius Kircher (1602-1680) was that dead flies in honey water produce "worms" by spontaneous generation. Was this a scientific statement? Some might say that it was not, because the statement turned out to be false. But that is not a fair argument, as it's made in retrospect and hence may be used against many current "scientific" statements in time. Others may say that Kircher's statement is un-scientific because his experiment was not controlled. But what does "control" mean? It means that some unwanted possibility is kept under control. However, unwanted possibilities have to be identified first, as we found out in the inductivism debate.

It was the Italian biologist and physician Francesco Redi (1621-1697) who came up with the idea that the honey-water only serves to attract living flies to drop their eggs into it. He had been inspired by William Harvey's claim that all life emerges from eggs—in defiance of spontaneous generation. Hence, Redi designed a controlled experiment with flasks sealed with a cork to prevent flies from depositing their eggs. Thus he could prove that "worms" do not come forth from rotting meat at all, let alone by spontaneous generation.

Nevertheless, Redi's experiment was not controlled in another way. He did not realize that sealed flasks do more than preventing eggs from coming in; they also prevent air from coming in, which then might also inhibit spontaneous generation. In other words, the theory of spontaneous generation was not really falsified as there was an uncontrolled variable involved. Therefore, Redi decided to replace the cork with a fine net in order to bring about a better controlled experiment more acceptable to his opponents. Although no "worms" appeared, rotting still occurred. In other words, the theory of spontaneous generation was actually confirmed instead and brought back to life.

In the next century, fresh evidence came from the English biologist John Needham (1713-1781). In 1745, he used a large variety of heated fluids containing small food particles in test tubes sealed with a cork. After a few days, he found tiny organisms in each test tube. Nowadays we tend to say that this finding was not scientific because something went wrong in his experiments. But what exactly went "wrong"? Isn't it "obvious" that something must have gone wrong, if you don't believe in the results?

Someone who did not believe in Needham's results was the Italian priest and physiologist Lazzaro Spallanzani (1729-1799). He had an idea of what went wrong in Needham's experiments and "improved" Needham's testing procedure by sealing the test tubes more thoroughly and heating them to a higher temperature. No "worms" appeared. Was this falsification of the theory of spontaneous generation? Not for Needham's camp; they replied that something had gone "wrong" in Spallanzani's experiments as well. The high temperatures must have weakened, or even destroyed, the "active principle" of spontaneous generation in the fluid.

Apparently, the discussion was still far from settled. It would take the genius of Louis Pasteur to put the last nail on the coffin of spontaneous generation (Case #5). Scientific findings and discussions are not always as conclusive as we would like them to be.

Case #3: Galileo – Orbits

For centuries, the orbits of planets had been explained by Ptolemy's geo-centric theory, which has all planets orbit around the earth. Based on this model, sailors had been able to navigate the seas with their ships, and astronomers had been able to predict eclipses. So that's not what the problem of geocentrism was. But its model had become such a complex system, with its numerous epicycles—circles moving on other circles—that hardly anyone believed it corresponded to the physical reality of the Universe. It worked all right but it was not believed to be true.

Here is the problem. When observed from one night to the next, a planet appears to move most of the time from West to East against the background stars. Occasionally, however, the planet's motion will appear to reverse direction, for a short time, from East to West. This reversal is known as retrograde motion. How had Ptolemy been able to explain this? He hypothesized that planets are attached, not to the concentric spheres themselves, but to circles attached to the concentric spheres. These circles were called "epicycles", and the concentric spheres to which they were attached were termed the "deferents." Then, the centers of the epicycles executed uniform circular motion as they went around the deferent at uniform angular velocity, while at the same time the epicycles (to which the planets were attached) executed their own uniform circular motion. That's how retrograde could be explained! Yes, but at the cost of being very convoluted.

It was time for a change, which came in 1543 when Nicolaus Copernicus (1473-1543) launched his helio-centric theory saying that the earth travels around the sun, as do the other planets. Copernicus explained the apparent retrograde motion of the planets as a result of the relative speeds of the earth and the planets circling around the Sun as observed from the earth. Thus, he was able to explain retrograde motion in a way and with a theory different from Ptolemy's. A century later, Galileo Galilei (1564-1642) would adopt the Copernican model and passionately promote it as a proven theory.

Interestingly enough, whereas Ptolemy was able to calibrate his geo-centric model by adjusting the size and speed of the epicycles and deferents, Copernicus and Galileo were able to calibrate their helio-centric model by adjusting the size and speed of a planet's orbit. As to whether this makes the helio-centric model more scientific is debatable. Besides, the new model worked with circular orbits—instead of the more accurate elliptical orbits—so it would still need the so despised epicycles of the Ptolemaic model to correct for its own inaccuracy.

In addition, the helio-centric theory had its own problems and had to face several challenges—potential falsifications, that is. The most important issue was an argument that had been made nearly two thousand years earlier by Aristotle himself. If the Earth did orbit the Sun, the philosopher wrote, then stellar parallaxes would be observable in the sky. In other words, there would be a shift in position of a star observed from the earth on one side of the Sun, and then six months later from the other side.

When another astronomer at Galileo's time, Tycho Brahe (1546-1601), tried to detect the test implication of a parallax with his instruments—which were the most sensitive and accurate ones in existence at the time—he failed. Therefore, he concluded the helio-centric theory had been falsified. So he came up with an alternative theory, which had the Moon and the Sun revolve around the Earth, but had the other planets

(Mercury, Venus, Mars, Jupiter, and Saturn) revolve around the Sun; so he had the Sun with those planets together revolve around the Earth. His model was a combination of geo-centrism and helio-centrism. From a strictly mathematical point of view, the two models were equivalent—both could be used to predict the motions of the planets with great accuracy. Tycho's model could explain also the phases of Venus. It is certainly not unusual that there may be several hypotheses to explain the same phenomena.

Another potential falsifier of the helio-centric model was the so-called "tower argument." If a stone is dropped from the top of a tower erected on the moving earth, it will fall towards the center of the earth. While it is doing so, the tower will be sharing the motion of the earth, due to its spinning. By the time the stone reaches the surface of the earth, the tower will have moved around from its original position and therefore, the stone would strike the ground some distance from the foot of the tower. But, in fact, the stone strikes the ground at the base of the tower. So the earth cannot be spinning. Right?

No, not quite! Galileo did not accept this potential falsi-fication. Instead, he developed the concept of relative motion and the law of inertia. An object held at the top of a tower and sharing with the tower a circular motion around the earth's center will, after it is dropped, continue in that motion along with that tower and will strike the ground at the foot of the tower. This way, Galileo could avoid falsification by giving a different interpretation to his observations.

Case #4: Mendel – Genes

Gregor Mendel (1822-1884) is often considered the "father" of genetics. It is not quite clear whether that is entirely correct. At first sight, it looks as if Mendel is really talking genetics. He did his experiments with garden peas and used seven different pairs of contrasting traits—such as seed shape (round vs. wrinkled) and stem length (short vs. tall)—to find out how these traits were "inherited."

This is probably something he had learned from plant-breeders. When plants with round peas were crossed with plants having wrinkled peas, all the offspring produced round seeds. Because one trait seemed to "dominate" over the other trait, Mendel called such traits *dominant*. When he let these new plants pollinate themselves, he found that 75% produced round seeds, but 25% wrinkled seeds again—which is a ratio of 3:1. In other words, the characteristic of wrinkled seeds had receded only temporarily, so he called that trait *recessive*. Doesn't all of this sound like modern-day genetics?

Perhaps not. It seems much more likely that Mendel was not really interested in the question of how the transmission of hereditary factors is achieved. Nowadays, we would say he was part of a paradigm, or research program, different from modern genetics. The paradigm of genetics centers on the transmission of hereditary factors—currently called genes. But that issue did not seem to be Mendel's main interest. Mendel was most likely still part of an older research tradition dating back to Carl Linnaeus (1707-1778) which centered on the question as to whether hybrids are really able to form a new species.

How do we know Mendel was not really talking genetics in his work? True, he comes deceivingly close to saying *AA* x *aa* produces 100% *Aa*, whereas *Aa* x *Aa* produces 25% *AA*, 25% *aa*, and 50% *Aa*. One might think that *AA* and *aa* are *homozygotes*, whereas *Aa* organisms should be called *heterozygotes*. But that impression is actually incorrect. In fact, Mendel symbolizes constant forms—which we would now call homozygotes—by just *one* letter (*A* or *a* instead of *AA* or *aa*). On the other hand, hybrids—which we would now call heterozygotes—he did express in our "modern" notation using *two* letters (*Aa*).

It is very important to notice that Mendel is very consistent in characterizing constant "forms" with *one* letter, and only the hybrids (as not being constant) with *two* letters. In Mendel 's view, an *Aa*-organism is a hybrid, not a heterozygote. A hybrid is a breed of two pure types and carries two different "elements" from each type. Even germ cells he described as "of form A."

In other words, Mendel's concept of hybrid seems to be part of the hybridization theories of the 19th century, and not of the genetic theories of the 20th century. That's what concepts do in science. Inspired by what species-breeders did, Mendel found out in his experiments that hybrids do not form a new species. When he crossed the two hybrids coming from two different pure forms (*AA* x *aa*) with each other (*Aa* x *Aa*), the two original pure forms could show up again. In other words, hybrids between two species are not constant new species but could be returned to the parental species by continuous backcrossing. This could be seen as a falsification of the hypothesis that hybrids form a stable new species. But that's not quite genetics.

How then could Mendel become the "father" of genetics? For almost 35 years, Mendel's findings escaped notice, until similar observations were made by "real" geneticists, who were working within the paradigm or research program of genetics—geneticists such as Hugo De Vries in Holland, Erich Von Tschermak in Austria, Carl Correns in Germany, and William Bateson in England. When they came across Mendel's original paper from 1866, this helped them interpret their own data better. Correns and Bateson, in particular, were fair enough to pay homage to Mendel, and because of this, he has been coined the founder of what became known as "Mendelism." But the fact remains that for someone like Bateson, an *Aa* organism is a heterozygote, a genetic mixture of two different alleles, whereas for Mendel, an *Aa* organism is a hybrid, a breed of two types. What is in a word? Sometimes a complete conceptual revolution!

Another concept that went through a conceptual revolution was the concept of gene. Although Mendel did not speak of "genes" but of "elements," his elements could be interpreted by his successors as "genes." One of Mendel's first hypotheses states that two different elements in a hybrid plant (of which one is dominant and one recessive) segregate and pass equally into the progeny. If so, self-pollination is supposed to lead to a ratio of 3:1, with 75% showing the dominant feature and 25% the recessive feature. The fact that in crossing plants with long and short stems, Mendel actually obtained a ratio of 2.84 : 1 was not reason enough for him to revise his hypothesis. In this case the observation needed to be corrected and interpreted within the setting of a more or less implicit statistical theory concerning sample size and sampling errors.

Later on, Mendel's successors found out that the purple color of snapdragon flowers, for instance, did not behave according to a 3:1 ratio of purple and red flowers, but rather according to a ratio of 9:3:3:1, namely 9 purple, 3 red, and 4 white flowers. Nevertheless, they did not accept this as falsification but instead maintained Mendel's hypothesis and found a solution by adapting some boundary conditions. Their escape clause was that this phenomenon had to be based on two genes instead of one, namely one gene for the formation of the color anthocyanin and one gene determining the pH (anthocyanin is purple in an alkaline surrounding, but red in acid). This move turned out to be successful.

Things began to change further when geneticists discovered that two or more pairs of genes may not *always* separate independently as the earliest geneticists had postulated. Sometimes genes seem to be *linked* to each other. It had already been noticed by the geneticist Walter Sutton that genes always occur in pairs and that chromosomes also occur in pairs. Was this mere coincidence? Or could it be that genes and chromosomes are connected, and that genes are actually located in chromosomes? The latter possibility seemed to be an attractive hypothesis. Humans, for instance, have only 23 pairs of chromosomes, but they have thousands of different genes. If there is a connection between genes and chromosomes, then there must be many different genes per chromosome. If so, genes located on the same chromosome may not separate independently but must be "linked," whereas genes located on different chromosomes can still separate independently of each other.

What we see happing here in the development of genetics is that genes gradually became more and more concrete, material entities. They evolved from the abstract hypothetical "elements" Mendel spoke of, to the more concrete "factors" Bateson referred to, then to what Johansson called "genes," and finally to the "beads" on Morgan's chromosomes. Yet, for a long time, they were still taken more as accounting or calculating units than as material entities. This changed drastically when James Watson and Francis Crick introduced their DNA model in 1953 (case #10). Since then, our understanding of the material basis of the gene has gone through an accelerating growth process.

Yet, for a while, acceptance of chromosomes as carriers of genes remained controversial. The geneticist Thomas Hunt Morgan initially rejected the chromosome theory of heredity, but then embraced it wholeheartedly in his classic 1910 paper in the journal *Science*. At last, Mendel's hypothetical elements had found a material basis. The chromosome theory could at least explain why the gene model works. And the concept of gene kept evolving with it.

Case #5: Semmelweis – Fever

The story of what Ignaz Semmelweis (1844-1848) discovered is a classic example of how evidence may be staring us in the face without being understood. It shows how observations easily elude us if we don't have the right concepts available. The Semmelweis case clearly demonstrates that "seeing" in science is more than observing, registering, or recording in a neutral way.

Here is what happened before Semmelweis entered the scene. Until the late 1800s, surgeons of the time did not scrub up before surgery or even wash their hands between helping patients, thus causing infections to be transferred from one patient to another, without anyone even knowing it. Doctors and medical students in the maternity wards, for instance, routinely moved from dissecting corpses to examining new mothers without first washing their hands, thus causing death by "puerperal" or "childbed" fever as a consequence. As anatomical, pathological dissection became more important to medical practice in the 1800s, the risk of infecting new mothers only got bigger.

Strange as it may sound, the idea of surgeons washing their hands is only 150 years old. Until then, all hospitals were pools of filth. Surgeons loved to speak of the "good old surgical stink" and took pride in the stains on their unwashed operating gowns as a display of their experience. The man who finally discovered the need for surgeons, and others, to wash their hands was Ignaz Semmelweis.

Semmelweis was a physician at a hospital in Vienna. The hospital housed two obstetric clinics, the first for teaching medical students, the second for training midwives. Semmelweis soon discovered an odd statistic in the records: maternal mortality rates at the first clinic, where the students were trained, had twice or sometimes three times the rate of the midwife-staffed second clinic. To explain the difference between the two clinics, Semmelweis tried eliminating various possibilities, ranging from the position of giving birth to eliminating a walk-through by a priest when patients were dying. But he could not find any confirming evidence. He excluded "overcrowding" as a cause, since the Second Clinic was always more crowded, and yet the mortality rate was lower. Other people had assumed the cause to be an epidemic, but Semmelweis found this falsified by the fact that fever hardly occurred in the city of Vienna itself.

After eliminating various other hypothetical variables, Semmelweis came to the conclusion that the only *significant* difference between the two clinics had to be the staff: department I was run by medical students, department II by midwives. Wasn't that an obvious difference? Apparently not, until Semmelweis hypothesized that an examination by students might be rougher than usual. However, the replacement of some students by midwives was not successful. Early in 1847, by "chance," Semmelweis noticed that a colleague wounded by the lancet of a student while doing an autopsy, died with the same symptoms that Semmelweis had observed in the victims of child-bed fever. As said earlier, chance observations only favor prepared minds. Semmelweis had a prepared mind.

This led Semmelweis to note that the doctors and medical students often also performed autopsies, while the midwives did not. It turned out to be a rather simple conclusion. The medical students took no hygienic steps between working on autopsies and delivering babies, and often used the same instruments for both activities. So he hypothesized that "particles" from the cadavers were responsible for transmitting the disease. He actually spoke of "cadaveric matter," which the students' hands carried from the corpses under investigation to the women under treatment. It may look like a strange concept, but it did throw some light on the problem under investigation.

Based on this concept, Semmelweis reasoned that child-bed fever could be prevented by destroying the cadaveric material adhering to hands. He therefore issued an order requiring all medical students to wash their hands in a solution of chlorinated lime. And promptly, the mortality from childbed fever began to decline. But soon after, Semmelweis broadened his hypothesis. When he had examined a woman with cervical cancer and then moved on to examine twelve other women in the same room, he noticed that eleven of them would die from childbed fever. So he concluded that the cause of childbed fever was not only "cadaveric matter," but also "putrid matter derived from living organisms." Finally, Semmelweis had "seen" what caused child-bed fever.

Unfortunately for Semmelweis, the medical community would persist in ignoring his findings and even ridiculing them. The main problem for Semmelweis was that his results lacked scientific explanation at the time. That became only possible some decades later, when Louis Pasteur (case #6), Joseph Lister, and others developed the germ theory of disease. Only then did Semmelweis' practice earn widespread acceptance, years after his death, when Joseph Lister, acting on Louis Pasteur's confirmed germ theory, practiced and operated with great success, using hygienic methods of washing hands and sterilizing equipment. But even today, convincing health care providers that they should wash their hands often remains a challenge.

Case #6: Pasteur – Germs

The final blow for the theory of spontaneous generation (case #2) came from the French microbiologist Louis Pasteur

(1822-1895). He proved through falsification that life doesn't emerge from spontaneous generation. Instead he confirmed William Harvey's hypothesis that all life emerges from life ("*omne vivum ex vivo*").

Pasteur was already familiar with micro-organisms. He had demonstrated that a tiny microorganism in yeast is responsible for the fermentation of sugar into alcohol. He also demonstrated that, when a different microorganism contaminates the wine, lactic acid is produced, making the wine sour. This led Pasteur to the idea that micro-organisms causing fermentation could also be the cause of certain diseases. Thanks to this knowledge and to the concept of micro-organisms, he was able to falsify the idea that life emerges from spontaneous generation.

He achieved this by using flasks with swan-necks drawn out in a flame, so dust particles and germs would get trapped in the lower curve; plus he used a yeast broth heated at high temperatures, assuming this would kill all micro-organisms. After several weeks, Pasteur observed that the broth in the swan-neck flask had not changed at all; but when the swan neck was broken, germs readily entered the flask and made the broth look discolored and cloudy with molds and micro-organisms.

Pasteur thus showed two things to the supporters of spontaneous generation. First, although the fluid was boiled, he could show that heating could not have destroyed any "active principle," since the fluid still had the ability to support life when the swan-neck was broken. Second, he had not barred any "active principle" from the air—as his opponents had suggested—because air was still free to enter and leave through the unobstructed pathway of the neck.

The outcome seemed to be devastating for the theory of spontaneous generation. But the dispute was not quite settled yet. The theory of spontaneous generation was still having some strong supporters. Between 1861 and 1863, Felix Pouchet carried out the same experiments as Louis Pasteur. They both used flasks with necks drawn out in a flame; they both used a nutrient solution, boiled and sealed. But there were also some "minor" differences. Pasteur used a yeast infusion, Pouchet a hay infusion. High in the mountains, the neck of the flasks was broken; to do so, Pasteur used heated pincers, Pouchet a heated file to break the neck. Pasteur found no rotting in 19 out of 20 flasks; Pouchet, on the-other hand, found rotting in 8 flasks out of 8.

As said before, sometimes falsification is not what it seems to be. What had gone "wrong" this time? Pasteur blamed Pouchet for using a file instead of a forceps. Pouchet, on the other hand, could have blamed Pasteur (but did not) for not having used a hay infusion. Had Pasteur used a hay infusion, he would have been in for quite a surprise, as it is known since 1876 that hay infusions support a spore that is not easily killed by boiling.

Nevertheless, a very partial commission took Pasteur's side in the debate. Hence, Pasteur with the help of the commission received the honor of having dealt a final blow to the theory of the spontaneous generation of life. The debate was settled. The most up-to-date scientific statement read: "Where life is absent, life cannot arise." Pasteur's experiment had now been widely accepted as conclusive and final.

But what is "final" after all? At the beginning of the twentieth century, Henry Bastian was going to revive the theory of spontaneous generation, when he unknowingly hit upon heat-resistant bacteria spores. And in 1953 Stanley Miller attempted to demonstrate how life could possibly arise in a prebiotic atmosphere. Miller's successors have been rather successful since. Now that the debate had been settled, it is easy to claim at hindsight that many of the earlier statements on spontaneous generation turned out to be false and therefore, should not be called scientific. But that is too easy an answer. Serious scientists had been involved in the debate. What was said then was based on what one knew at that point in time. Performing experiments "correctly" is not just a matter of methods, but also of theories.

The main issue at hand in the dispute between Pasteur and Pouchet was: what happens when a nutritive medium, which is sterilized by boiling, is exposed to clean air? Pasteur protected his theory by defining all air that gave rise to life in his boiled flasks as contaminated. This was a decision he made. Notice the assumptions he held: germs can be killed by heat and germs can contaminate air. With the latter assumption, he refused to be swayed by what, on the face of it, was falsifying evidence. Or should we rather say that he was obstinate in the face of falsifying evidence? It is only afterwards that it was possible to designate his decision as successful.

Case #7: Darwin – Evolution

Charles Darwin's theory basically asserts that all living entities have a common origin, in spite of apparent differences—so-called "common descent with modification." In his 1859 book *On the Origin of Species*, he elaborated extensively on various kinds of evidence for common descent. He spoke of a "theory of descent with slow and slight successive modifications." The modification part he explained by using a new concept, "natural selection." After launching his famous theory of natural selection, Darwin not only introduced a (partially) new theory but also a new research program with a hard core and a protective belt.

On the one hand, he protected his theory against what might be taken by some as falsifying evidence. Darwin knew very well his theory of natural selection could face much counterevidence, but he had the feeling that these objections should not falsify the "hard core" of his fledgling theory. That is why he tried to provisionally invalidate them in chapter six of his 1859 book *The Origin of Species*. In that chapter, with the title "Difficulties of the Theory," he discussed problems such as "missing links" in the paleontological records, the imperfection of certain biological features, and so on.

On the other hand, Darwin's theory opened the way for experiments designed to test the "hard core" of his program. If natural selection indeed promotes good designs over bad designs—which supposedly increases their survival rates and thus their frequency in later generations—all of that can be tested. Put differently, natural selection "selects" by favoring causes that have "successful" effects. This makes for the "*positive* heuristics" of his research program instructing scientists which paths of research to *pursue*. It shows the explanatory power of natural selection—without claiming, though, that natural selection is the only factor operational in evolution.

Many biologists took on this task. Bernard Kettlewell began to study how the composition of the peppered moth population experienced a change when industrial areas in England became more polluted. H.W. Bates started the study of mimicry, the phenomenon that an unprotected species (say, certain flies) takes advantage of its partial resemblance to a protected species (say, certain wasps), and then Bates discovered that when the protected species varied geographically, its mimicking satellites had undergone exactly the same changes as their unpalatable models. And then there was Georges Teissier who designed simple population cages to experiment with different kinds of selection pressure on easy and quickly to breed fruit flies.

Soon after, we see population geneticists using gene-pool models that they subjected to specific selection pressures.

Darwin's research program was so powerful that many biologists nowadays are still successfully working within the hard core of its framework. New channels for fruitful experiments had been opened, which would lead to corrections of the theory without affecting its hard core. This has eventually led to Neo-Darwinism with its so-called "synthetic" version of the theory of evolution.

One of the main challenges of Darwin's theory was the confrontation with another paradigm prevalent before Charles Darwin, which hypothesized a perfect adaptation of organisms to their environment under a Creator's design. In a world of instant designing, non-adaptations are anomalies, but in a world of natural selection, they are not, because a trait that was formerly beneficial may still be inherited while no longer beneficial. Think of features like these: the wings of ostriches and penguins, although they no longer fly, or the eyes of bats, although they no longer use them, or the webbed feet of upland ducks which only rarely go near the water.

Before Darwin, such anomalies were hard to explain—or even went unnoticed. But Darwin's theory could explain them as a result of natural selection which promotes the best designs available at the time—which are usually not perfect designs. They can even become outdated designs which are no longer functional and then become "rudimentary"—that is, "left-over" parts from common descent, because natural selection works under constantly changing circumstances. Rudimentary features only make "sense," if we assume that all organisms derive their features from common ancestors. That's the hard core of Darwin's theory.

Case #8: Fleming – Antibiotics

Alexander Fleming (1881-1955) had always been searching

for new inhibitors of bacterial growth. While nursing a cold in 1921, Fleming discovered lysozyme—a mildly antiseptic enzyme present in body fluids—when a drop of mucus dripped from his nose onto a culture of bacteria. Thinking that his mucus might have some kind of effect on bacterial growth, he mixed it with the culture. A few weeks later, he observed that the bacteria had been dissolved. Always on the lookout for natural bacteria killers, this observation excited Fleming enormously. It marked Fleming's first great discovery, preparing him for a much bigger one.

In his own words, "The view has been generally held that the function of tears, saliva and sputum, so far as infections are concerned, was to rid the body of microbes by mechanically washing them away... however, it is quite clear that these secretions, together with most of the tissues of the body, have the property of destroying microbes to a very high degree." What was guiding him in his research was the *concept* of a bacterial growth inhibitor.

Now Fleming was ready for his biggest discovery. It happened in 1928. He was a brilliant researcher, but his laboratory was often untidy. Before leaving for vacation, he had stacked all his cultures of staphylococci on a bench in a corner of his laboratory. While he was away, one of his assistants had left a window open and the dishes had become contaminated by different microbes. When Fleming returned to his laboratory and looked at the plates, he noticed that one culture was contaminated with a fungus, and that the colonies of staphylococci immediately surrounding the fungus had been destroyed, whereas other staphylococci colonies farther away were normal. This made him famously remark, "That's funny." Then he showed the dish to an assistant, who remarked on how similar this seemed to Fleming's famous discovery of lysozyme.

Fleming decided to grow the mold in a pure culture and

found that it produced a substance that killed a number of disease-causing bacteria. He identified the mold as being from the genus *Penicillium*, and, after some months of calling it "mold juice," renamed the substance it released "penicillin" on 7 March 1929.

Although his predecessors had already been cautious about molds in their bacterial cultures, they had never made the observation Fleming made. The reason is simple: they were not searching for something designated by the concept of growth inhibitors—and that is why they were unable to find penicillin. Just opening one's eyes is not enough to make discoveries based on observations. Usually searching precedes finding. It took someone like Fleming to notice on his dishes of bacterial cultures a zone around an invading fungus where the bacteria could not seem to grow. Suddenly, this became an observation that made sense. Thinking he had found an enzyme more powerful than lysozyme, Fleming decided to investigate further. What he found out, though, was that it was not an enzyme at all, but an antibiotic—one of the first antibiotics to be discovered.

Was it mere luck? In Fleming's own words: "Some bacteriologists probably noticed similar changes [...] but their cultures had been discarded, because there was no special interest in natural compounds having an anti-bacterial effect. [...] Fortunately, however, I had always been searching for new inhibitors of bacterial growth." Fleming happened to be in search of such compounds before he could find them. But it was certainly not mere "luck." He was a prepared mind who had the concept of growth inhibitors to his avail.

Case #9: Le Verrier – New Planets

Most astronomers at the time of the French astrophysicist

Urbain Le Verrier (1811-1877) worked within the framework of Isaac Newton's research program. Newton had been able to combine four laws—those governing the relations between the forces acting on a body and the motion of that body—into a coherent deductive system in which all of the principles of mechanics could be deduced as theorems from only four axioms. These axioms were (1) the law of inertia, (2) the law of force, (3) the law of action and reaction, and (4) the law of gravitation varying with the square of the distance ($1/d^2$). From these four axioms Newton was able to derive the laws of planetary orbits, falling bodies, projectile motion, and the variations of the tides. Newton demonstrated that Kepler's laws and Galileo's law of free fall could be derived from his four axioms. Moreover, he achieved a unification of astronomy and mechanics.

Astronomers loved to work within this research program. After the planet Uranus was discovered in 1781 by William and Caroline Herschel, astronomers were able to predict its orbit by using Isaac Newton's unified laws. However, when they carefully followed Uranus' motion in its slow 84-year orbit around the Sun, they began to notice that something was wrong. Uranus didn't quite move as it should according to Newton's theory. So they came up with *ad-hoc* adjustments— they refined their measurements, took more and more careful observations—yet the anomaly didn't go away.

Obviously, astronomers of the day didn't think the unexpected observation falsified Newtonian gravity. Instead, they came up with another hypothesis for the strange motion of Uranus by postulating some large and unseen object that "must" have affected the planet's orbit. Calculations showed that it would have to be a planet as large as Uranus and even farther away from the Sun.

It was not until 1846 that the French astrophysicist Urbain

Le Verrier predicted with the use of Newton's laws where this hypothetical planet had to be located. Some of his colleagues in Germany had the courage to direct their telescopes to the spot where Le Verrier had told them to look. Lo and behold, within half an hour, they spotted the planet Neptune. Instead of a demise, this observation was a victory of Newtonian physics.

Next, Le Verrier went after another planetary puzzle. Several years after his discovery of Neptune, it became clear to him and other astronomers that Mercury was another planet that wasn't moving either as it was supposed to. The point in its orbit where it made its closest approach to the Sun shifted a little more than Newton's gravity said it should each Mercurial year. Just as with Uranus before, the anomaly didn't go away with more persistent observation. It stubbornly remained, defying what Newton had claimed. Another falsification?

No, once again, Newtonian gravity was not thrown out as falsified—at least, not immediately. Instead, Le Verrier tried the same stunt again by calling in an unseen planet, a tiny rock so close to the Sun that it must have been missed by all other astronomers. He called the planet Vulcan. Le Verrier and others feverously searched for Vulcan, aiming powerful telescopes to solar eclipses in an attempt to catch a glimpse of the unseen rock during the brief minutes while the Sun was totally blocked by the Earth's moon.

No one was ever able to spot Vulcan. The astronomy community finally gave up the search, concluding that Vulcan simply wasn't there—another invention that did not make it to a discovery. But even so, Newton's theory of gravitation wasn't discarded. For years, the mystery of Mercury's orbit remained unsolved, without any serious suggestion that Newton's theory was wrong. Falsification was simply not on anyone's mind... until finally, in 1915, Albert Einstein used his bold and daring theory of general relativity to show that he could succeed where

Le Verrier had failed.

Einstein's theory of general relativity was a new hypothesis of how gravity worked, more accurate than Newtonian physics. It perfectly predicted the shift in the point of Mercury's orbit closest to the Sun. Einstein said he was "beside himself with joy" when he was able to confirm that his theory could correctly solve the longstanding puzzle of Mercury's orbit. Four years later, the British astronomer Arthur Eddington and his team aimed their powerful telescopes at an eclipse to confirm that starlight does bend around the Sun as Einstein's theory had predicted. Einstein was instantly rocketed to fame as the man who had shown Newton wrong.

So Newtonian gravity was ultimately thrown out, but not just because of observations that had threatened it. It wasn't until a viable alternative theory had arrived, in the form of Einstein's general relativity, that the scientific community could accept the notion that Newton might have missed a few things. The general thought about this outcome is that Newton's theory, which had been more comprehensive than Galileo's or Kepler's by explaining more phenomena, had now in turn been subsumed under, or reduced to, Einstein's theory. But is this really a case of expansion?

It is very tempting to explain the transitions between theories as a case of an older theory being "absorbed" by a newer one. But that's not quite true. Galileo's law of falling bodies, for instance, cannot be deduced from Newton's axioms, not even in conjunction with the additional premise that the falling body is "close" to the earth. According to Galileo, a body accelerates constantly as it falls to the earth, but according to Newton the acceleration is inversely proportional to the square of the distance from the center of the earth, and thus increases steadily as the body falls. Thus, Newton's theory does not absorb or contain Galileo's law. This also holds for the relation between Kepler's laws and Newton's theory. And with Einstein's theory of general relativity the same has happened to Newton's laws.

In other words, Einstein's theory did not nicely absorb Newton's theory, and Newton's theory did not nicely absorb the laws of Kepler and Galileo. In the course of history, they supersede each other, without nicely absorbing each other. There is not always a smooth succession or accumulation of theories. Very often the process goes in fits and starts.

Case #10: Watson and Crick – DNA

In 1952, James Watson and Francis Crick drafted a research program which was in for a successful confirmation. Scientists before them had handed them important, useful, and relevant information. Erwin Chargaff, an Austrian biochemist, was one of them. He had read the famous 1944 paper by Oswald Avery and his colleagues at Rockefeller University, which demonstrated that hereditary units, or genes, are composed of DNA. This paper had a profound impact on Chargaff, inspiring him to launch a research program that revolved around the chemistry of nucleic acids, which are the building blocks of DNA. Of Avery's work, Chargaff wrote the following: "This discovery, almost abruptly, appeared to foreshadow a chemistry of heredity and, moreover, made probable the nucleic acid character of the gene... Avery gave us the first text of a new language, or rather he showed us where to look for it. I resolved to search for this text."

Chargaff had noticed that in DNA the amount of adenine (A) is usually similar to the amount of thymine (T), and the amount of guanine (G) usually approximates the amount of cytosine (C). In other words, the total amount of purines (A+G) and the total amount of pyrimidines (C+T) are typically equal.

Chargaff's research would be vital to the later work of Watson and Crick, but Chargaff himself could not imagine the explanation of these relationships—specifically, that A is bound to T and C is bound to G within the molecular structure of DNA.

It was Chargaff's realization that A=T and C=G, combined with some crucially important X-ray crystallography work by English researchers Rosalind Franklin and Maurice Wilkins, that contributed to Watson and Crick's derivation of the three-dimensional, double-helical model for the structure of DNA. Watson and Crick's discovery was also made possible by recent advances in model building, or the assembly of possible three-dimensional structures based upon known molecular distances and bond angles, a technique advanced by American bio-chemist Linus Pauling. It's another case of realism in science (§14).

Using cardboard cutouts representing the individual chemical components of the four bases and other nucleotide subunits, Watson and Crick shifted molecules around on their desks, as though putting a puzzle together, which was done before computer simulations were possible. Watson and Crick were misled for a while by an erroneous understanding of how the different elements in T and G were configured. Only upon the suggestion of American scientist Jerry Donohue did Watson decide to make new cardboard cutouts of the two bases, to see if perhaps a different atomic configuration would make a difference. It did. Not only did the complementary bases now fit perfectly together (i.e., A with T and C with G), with each pair held together by hydrogen bonds, but the structure also nicely reflected Chargaff's observation that A=T and C=G.

Watson and Crick were quite positive that the distance between two successive nucleotides in a DNA-chain had to be related to the recurring distance of 0.34 nm. Furthermore, they assumed that the recurring distance of 2.0 nm was an indication of the chain's width. But what about the third recurring unit—a distance of 3.4 nm? A solution might be to coil the chain up like a helix; the regularity of 3.4 nm would thus be the distance between the successive coils of the helix. They arrived at a coil of ten nucleotides, because 3.4 nm equals ten times the distance between two nucleotides.

Further calculations showed that a single-stranded chain would have only half the density found in DNA, which gave them the idea of a *double*-stranded helix. A scale model helped them to realize several arrangements until a suitable candidate was found: two strands twisted in opposite direction around an imaginary cylinder, while its bases are turned inward.

As Watson and Crick were wrestling with the structure of DNA, they started to wonder how the base sequence of DNA could be translated into the amino acid sequence of proteins. Even before the meaning of the structure of DNA had been seen, Watson had already coined the hypothetical phrase "DNA makes RNA makes protein." They came up with an elegant model of DNA *transcription*, from DNA to RNA. The test implications of this hypothetical model were also confirmed by later experiments.

The DNA-model also helped them get hold of a good mechanism for DNA *replication*, from DNA to new DNA. In 1955, the test implications of this theory were tested and, again, confirmed. Soon a new hypothesis came up, positing that replication might be based on enzymes. In 1956 the first enzyme was identified. As a result, the double-stranded DNA-model received stronger and stronger confirmation and became so well established that the discovery of *single*-stranded DNA, found in bacteriophages in 1958, was not considered as counter-evidence that falsified their theory, but just as an exceptional variety of regular DNA. This did not affect the "hard core" of their research program, but only its "outer belt."

II

Science, and Only Science?

Because science has been so successful in answering so many questions about the world around us, it seems very likely that science may in time answer *all* our questions, if we could only be patient enough. Of course, that is not a logically valid argument, because it is of the inductivist type. Yet, it seems to be an attractive line of reasoning. But can it be true? We will see soon why it can't be.

We could mitigate this wide-ranging claim by saying instead that science may in time have an answer to all our *scientific* questions. But even that claim seems questionable. Although this is basically an empirical question—only time can tell—there may be reasons why science can never deliver all the final (scientific) truths about this Universe. The problem here is that the extent of what is still unknown is in itself unknown. It is impossible, just on philosophical grounds, to know that which is still unknown. Besides, how would we ever know there are no more unanswered scientific questions left? Absence of evidence for more unanswered questions is not evidence of their absence.

5. The Megalomania of Some Scientists

So far we have seen what science *can* do for us. It seems to be quite a bit—enough to make many people think science can do anything else for us. As a matter of fact, the tremendous power of scientific reasoning demonstrated daily in the many marvels of modern technology has empowered some scientists to think science can do everything we could ever think or dream of. They consider science a know-all and cure-all, at least potentially so. They passionately air this conviction in their books, through the mass media, and over the internet. They are eager to shout their message through the megaphone of media and academia.

Here are some of their voices. The chemist and Nobel-Laureate Peter Atkins thinks that scientists are privileged "to see further into truth than any of their contemporaries... [He believes] there is no reason to expect that science cannot deal with any aspect of existence." Notice the word "any." The astrophysicist Carl Sagan said something similar, "Science for its part will test relentlessly every assumption about the human condition." Notice the word "every." The cosmologist Stephen Hawking once exclaimed, "[O]ur goal is a complete understanding of the events around us and of our own existence." The word "complete" stands out here. And then there is the very vocal biologist Richard Dawkins who thought he had good reason to say, "[G]aps shrink as science advances, and God is threatened with eventually having nothing to do and nowhere to hide." Dawkins pretends to speak on behalf of many of his colleagues, expressing what he believes science has shown us— that eventually God has nothing to do anymore.

Undoubtedly, science has a great track record. But so does Hollywood. Having a great track record does not make something worth *every*-thing. Nevertheless, it remains a timeless temptation to make extravagant claims about science. Science and scientific knowledge have thus been endowed with the status of a new, comprehensive philosophy of life, close to a new religion. The "building" that houses science has become a "temple" of science. Many still see this image in the way the Massachusetts Institute of Technology (MIT) was built in the early 1900s—with its impressive pillars. The University of Cambridge in England may not look like a temple, but it does treat its members Hawking and Dawkins as little "gods."

Admittedly, it is true that if science does not go to its limits, it should be considered a failure, but it is equally true that, as soon as science oversteps its limits, it becomes arrogant—a know-it-all, based on a form of gross megalomania. That's why it is a constant temptation for scientists to think that their expertise in scientific research makes them automatically experts in everything else in life. However, the scientists we quoted at the beginning of this chapter fail to remember they are just specialists like any other specialists; they are specialists in doing scientific research regarding the material aspects of this world—physical, chemical, biological, or whatever—leaving everything else for other "specialists" to deal with.

If scientists claim expertise in everything else than science as well, they are like plumbers trying to also fix our electricity at home—or like electricians attempting to fix our plumbing. To claim that science provides the one-and-only source of reliable knowledge is an authoritarian ideology that has become known as *scientism* (§6)—a glorious word for some, a dirty word for others.

But scientists are not just specialists in the broad sense of doing scientific research. They have become even more "specialized" in specific fields and branches of science—actually becoming so specialized that the late Nobel Laureate and biologist Konrad Lorenz felt he could say that a scientist "knows more and more about less and less and finally knows everything about nothing." What comes with specialization is that physics, for example, discloses only one of the many aspects of the material world, its physical aspect, thus providing physical answers to physical questions phrased in terms of physical causes and effects. In other words, in physics we are in a realm different from that which is covered by biology. Biology, for one, offers another view of the Universe, providing biological answers to biological questions in terms of biochemical causes and biological functions. That's how "specialized" the various branches of science have become.

What also tends to come with specialization within the scientific community is the feeling that one's own field is superior over other fields. It's another form of megalomania. We find it, for instance, among some biologists who think that everything about life can be explained in terms of evolution— including the rationality and morality of the human world. An ideology of superiority and megalomania like this is usually called *evolutionism*—merely a more constricted form of scientism. It goes far beyond what evolutionary theory can claim.

But this form of megalomania is not only limited to biologists. It is even more pronounced among physicists who think nothing in science is really explained if it hasn't been reduced to physical explanations. But we should wonder whether there is really any need or desire to ultimately reduce all scientific phenomena to quantum events. One could argue instead that life and living phenomena cannot be reduced to the simple principles physics has identified without losing the essence of life itself. In order to explain a biological feature in physical terms, it must first be "physicalized," which is likely to sacrifice precisely the feature that gives it a biologically, or otherwise, distinctive character. As the chemist Arthur Robinson once said: "Using physics and chemistry to model biology is like using Lego blocks to model the World Trade Center." The instrument may simply be too crude.

Yet, physics is promoted by some as the one-and-only way of understanding the world. An ideology of megalomania like this one is usually called *physicalism*. It leads easily to a feeling of superiority. It is said, for instance, that the Los Alamos nuclear physicists during the Manhattan Project refused to consult medical doctors, but instead diagnosed themselves by reading medical books on their own, assuming that medical science must be trivially easy for anyone who could master nuclear physics. But they were not the only ones. Although, for a while, science in general was supposed to be of a better quality than non-science, nowadays certain parts of science are considered intrinsically better than other parts. Many scientists promote their own way of doing science as the single right way to do science at all.

What unites "-ism" ideologies such as physicalism and evolutionism is the more general ideology of *scientism* mentioned at the beginning of this chapter. What unites all scientists—such as physicists, chemist, biologists, neuro-scientists—is that they are all "specialists" in measuring and counting, by using the empirical cycle (§4). This emphasis on quantification has often led to a "mathematicalization" of science, reducing everything to mathematical equations and formulas. There is nothing wrong with that, as long as we realize that mathematical structure by itself is a mere abstraction; it cannot be all there is to it, because structure presupposes something concrete that *has* the structure. Albert Einstein, for one, always resented this trend of "mathe-maticalization" in physics. He did not want to be a slave to mathematics. For instance, after he had studied Fr. Lemaître's revolutionary 1927 paper intensely (§18), he told the priest, "Your mathematics is perfect, but your physics is abominable." Einstein would one day take those words back, but his point was that mathematics does not and should not have the last word in science.

The mathematician and Nobel Prize laureate Bertrand Russell once said, "Physics is mathematical, not because we know so much about the physical world, but because we know so little." Another mathematician, Alfred North Whitehead, called this truncated view the "fallacy of misplaced concreteness"—a fallacy that we commit when we mistake our abstractions for concrete realities. Niels Bohr once famously acknowledged, "If we want to say anything at all about nature— and what else does science try to do?—we must somehow pass from mathematical to everyday language."

This doesn't take away from the fact that scientists are known to be "specialists" in measuring, counting, dissecting, and testing. It is a specialty they usually are extremely proud of. It gives them a sense of superiority over people who don't have that expertise. Even those who are not scientists agree that what we know through science is much more reliable than what we know through non-scientific channels. There is this—often unspoken—assumption that the scientific method is not only the best method there is, but also the only method we have to understand the world.

This feeling of superiority makes many believe that all our questions have a scientific answer phrased in terms of particles, quantities, and equations. There is this strong belief that there is no other point of view than the "scientific" world-view— which is basically a dogmatic, unshakable belief in the omni-competence of science, suggesting there is no corner of the Universe, no dimension of reality, no feature of human existence beyond the reach of science.

Is that a justified belief? There is no reason to think so. Albert Einstein put this issue in the right perspective with a sign hung in his office at Princeton University that read, "Not everything that can be counted counts; not everything that counts can be counted." Limiting oneself exclusively to the quantitative aspects of this world and thus leaving other aspects out, as science does, does not make those other aspects disappear. There are so many questions science cannot possibly answer.

Science on its own does not and cannot say anything at all about things outside the material world of science—it does not and cannot even say that the material world of science is all there is—that would amount to *materialism*, another ideology. I know of a Jesuit biologist who used to tease his parishioners and challenge his students with a quip: "You don't need to tell me anything about life—I am a biologist." Coming from his mouth, it was a joke! But in the mouths of some scientists it is not.

This arrogance leads easily to slogans like "Life is nothing but chemistry and physics" (James Watson); "Humans are nothing but a speck of dust" (Carl Sagan); "Humans are nothing but a pack of neurons" (Francis Crick); "The brain is nothing but a 'meat machine'" (Marvin Minsky); "Humans are nothing but glorified animals" (Charles Darwin); or "Humans are nothing but a bundle of instincts" (Sigmund Freud); and the list goes on and on. In all these cases, the words "is" and "are" feature as keywords.

However, one could argue against claims like these that the best and most we can say is this: within the setting of a biological model, human beings may indeed be seen as "*only* molecules," but in reality they contain "*also* molecules." More specifically, human beings have *also* DNA, but in genetics they seem to be *only* DNA. Claiming that human beings are "nothing but" DNA ignores this important distinction.

C. S. Lewis famously dubbed this attitude "nothing-buttery." It is an ideology under the guise of science. It treats scientific models as if they are exact replicas of what they represent. Besides, "nothing-buttery" is basically suicidal. If indeed we were nothing but a "bag of molecules," then this very statement coming from a mere "bag of molecules" would not be worth more than its molecular origin, and neither would we ourselves who are making such a statement. Claims of "nothing-buttery" just defeat and destroy themselves. They cut off the very branch that the person who makes such claims is—or actually was—sitting on. This should put a science like biochemistry, for instance, in its proper place: it is a great specialty, but there must be more to life than molecules.

In other words, the following questions keep pressing: Is there really nothing more to life than biology? Is there really nothing more to this world than the quantum events of physics? Isaac Newton once quipped, "Gravitation is not responsible for people falling in love." People just don't "gravitate" towards each other that way. Love is one of those issues that science has only a very skimpy explanation for, because love is arguably more than a chemical reaction. As the Austrian physicist and Nobel Laureate Erwin Schrödinger once said about science, "It knows nothing of beautiful and ugly, good or bad, God and eternity. Science sometimes pretends to answer questions in these domains, but the answers are very often so silly that we are not inclined to take them seriously."

As a matter of fact, science cannot answer the many philosophical, moral, and theological questions people have come up with for centuries—questions such as "Why are we here?" or "What is truth?" or "What is love?" or "Is there free will?" or "What is good and what is evil?" or "Is there life after death?" or "Does God exist?" These are considered very legitimate questions by almost everyone. That these questions cannot be answered by counting and measuring doesn't mean that these questions and answers do not count. However, questions like these do have answers, although not scientific answers phrased in terms of particles, quantities, and equations. Does that make these questions and answers illegitimate? It is hard to see how they could.

Besides, there is this most fundamental question that science has no answer for—it's the famous question once worded by the philosopher (and mathematician) Gottfried Leibniz this way: *Why is there something rather than nothing?* Some scientists think science is able to answer that question by pointing at the laws of nature that science has discovered—gas laws, gravity laws, laws of thermodynamics, laws of genetics, and so on. These laws are supposed to explain why there is something rather than nothing. But that answer is hard, if not impossible, to accept, for these laws could have been other than they are, so they are not self-explanatory. Besides the question arises of why there are laws like these—rather than nothing? Physicist Paul C. Davies comments, "Over the years I have often asked my physicist colleagues why the laws of physics are what they are? ... The favorite reply is, 'There is no reason they are what they are—they just are.'" In one sense, his colleagues are perfectly right. They just *are*—there is no *scientific* explanation for the fact that the laws of nature *are* nor what they are.

Obviously, we are dealing here with a philosophical issue, not a scientific one. Science can never explain what science must take for granted. Even some scientists are aware of this philosophical problem. Since the laws of nature presuppose the very existence of the Universe, they cannot be used to explain the existence of the Universe. The astrophysicist Stephen Hawking, for instance, dares to ask this philosophical question: "What is it that breathes fire into the equations and makes a

Universe for them to describe? The usual scientific approach of science of constructing a mathematical model cannot answer the questions of why there should be a Universe for the model to describe. Why does the Universe go to all the bother of existing?" That was certainly a deep insight Hawking had.

The problem of answering this question is that laws of nature—whether physical, chemical, biological, or whatever—cannot explain their own existence. They could have been different, so they are contingent rather than necessary and thus could not provide an ultimate explanation of their own existence. Actually, *nothing* (no-thing) can explain how these laws came into existence. To use an analogy, you cannot be your own father or mother. You just cannot give your own existence to yourself or receive it from yourself. To think otherwise would create an inconsistency. Nothing can just pop itself into existence; it must have a cause, because it does not and cannot have the power to make itself exist. For something to create itself, or produce itself, it would have to exist before it came into existence—which is logically and philosophically impossible.

In other words, the search for causes—behind the existence of laws of nature and behind everything else in nature—must come to an end. There must be a cause that does not come into existence and is not itself in need of a cause. It is that latter necessary, transcendent cause that has become known, especially through the philosophy of St. Thomas Aquinas, as the *First Cause*. Only the First Cause can explain the existence of so-called *secondary* causes, which cannot explain themselves. It is only thanks to the First Cause that other causes can exist and become causes of their own and be the cause of other causes.

Science only deals with secondary causes. Even scientific laws such as the law of gravitation are only about secondary causes in terms of cause-and-effect relationships. But one could and should argue that all cause-and-effect connections we know of remain "hanging in the air" and miss any foundation, if the First Cause doesn't keep them in existence. One could also say that the First Cause provides a "point of suspension" for the chain of secondary causes itself, so to speak. The philosopher Michael Augros uses the simple example of an I-beam with a hook on it from which a chain is to be hung: "If there is nothing for that whole chain to hang from, it will not hang, and nothing can be hung from it. There is nothing about those links in themselves that makes them want to hang in space.... There must also be something *from* which things hang and which is not itself hanging from anything." In other words, without the necessary and eternal existence of a First Cause, there could not be a chain of the secondary causes with which science deals.

This is an important philosophical insight—rightly called a "proof of God's existence." It doesn't come from physics but from metaphysics, a branch of philosophy. This gives the proof strong deductively conclusive support, whereas science can only provide inductive evidence. It also sheds a powerful light on many questions we have. How, for instance, can we explain our own existence? We may go into biological causes, which we then can trace back to chemical causes, which in turn can be derived from physical causes. But all these causes combined don't really explain anything. For the entire chain of these causes is just floating in the air, and keeps doing so until we find a "foundation" for it to rest on, or a "hook" for it to hang on. That's why the existence of a First Cause is needed, which explains why all the other causes are only secondary causes in need of a First Cause.

Nonetheless, in spite of the previous powerful philosophical arguments, many scientists keep going much further than their science allows them to go. They have the conviction that science provides the one-and-only way of finding the only reliable knowledge there is. They have a strong conviction about what counts as "real" knowledge—and a very restrictive one at that. In short, they are devoted to scientism. Do they still have a foot to stand on? Let's find out.

6. What's Wrong with Scientism?

After explaining what science *can* do for us, we discussed briefly what science *cannot* do for us. Yet, scientism keeps maintaining that there is nothing that science is *not* able to do for us. Science has no bounds and no limitations according to the ideology of scientism. Can this be true? There are several, partly interconnected reasons why this cannot be true. Let's discuss the main ones—at least eleven in total.

A first reason for questioning the ideology of scientism is a very simple objection: those who defend scientism seem to be unaware of the fact that scientism itself does not follow its own rule. It cannot put itself to the test in an empirical or experimental way—the way scientific claims are supposed to be tested. But if so, then scientism would be like a boomerang coming back at the one who launched it. How could science ever prove all by itself that science is the only way of finding truth? There is no experiment and no empirical evidence that could do the trick. We cannot test scientism in the laboratory or with a double-blind experiment. Consequently, the truth of the statement "no statements are true unless they can be proven scientifically" cannot itself be proven scientifically. Science on its own cannot answer questions that are beyond the reach of its empirical and experimental techniques. Scientism is not a scientific discovery but at best a philosophical or metaphysical opinion—and a poor one at that. It declares everything outside science as a despicable form of metaphysics, in defiance of the fact that all those who reject metaphysics are in fact committing their own version of metaphysics. This makes scientism a totalitarian ideology, for it allows no room for anything but itself.

A second reason for questioning the ideology of scientism is that one cannot talk *about* science without stepping *outside* science. Stating that science is all there is, and that there can be nothing outside it, must be a claim made from outside the domain of science, thus grossly overstepping its own boundaries. By doing so, it loses the very scientific credentials it wants to defend. If scientism pretends to claim that there is nothing outside science and that there is no other point of view than the scientific view, it can only do so by doing what it's not allowed to do by its own verdict—namely, stepping outside science. So this step does not seem to be a very scientific move to validate its own statement. When scientism declares there is nothing outside the domain of science, it must be making a statement from outside the domain of science, which cannot be tested with tools and methods from inside the domain of science. Science cannot pull itself up by its own bootstraps—any more than an electric generator is able to run on its own power.

A third reason for rejecting scientism is that it discards very legitimate world-views and explanations that are not scientifically founded and validated. However, we may be able to neglect or ignore such explanations, but we cannot logically and validly conclude that therefore we must reject them as unreasonable or unfounded. Scientism basically dismisses anything that cannot be tested and corroborated by science, and in doing so it automatically rejects what it neglects. Something we neglect we cannot just reject, unless we have additional reasons to do so. If those additional reasons are missing, then there is still space left for world-views other than the scientific world-view. The late University of California at

Berkeley philosopher of science Paul Feyerabend, for instance, comes to the conclusion that "[S]cience should be taught as one view among many and not as the one and only road to truth and reality." Even the "positivistic" philosopher Gilbert Ryle expressed a similar view: "[T]he nuclear physicist, the theologian, the historian, the lyric poet and the man in the street produce very different, yet compatible and even complementary pictures of one and the same 'world.'" Science provides only one of these views, which makes promoting scientism a very restrictive, exclusive, and authoritarian move.

A fourth reason for rejecting scientism is that a methodology as successful as the one science provides does not disqualify any others. First scientism declares the methodology used in science as far superior to others, and then claims, in an unscientific way, that science has the only legitimate methodology that offers us the only reliable view on the world, thereby disqualifying any other views. A blood test, for instance, is an excellent method to assess a person's health. However, there are many other reliable methods, such as X-rays, MRIs, etc., depending on what we are trying to assess. But a blood test on its own cannot be used to prove that a blood test is the best, let alone only, method there is. Yet, that's what scientism does: first it declares science's empirical and experimental techniques as far superior to other methods, and then claims that this disqualifies any other methods than the ones science uses. Not surprisingly, scientism often results from hyper-specialized training coupled with a lack of exposure to other disciplines and methods. So it offers us a rather curtailed approach, certainly not a complete one.

A fifth argument against scientism is its claim that what science reveals us is *all* there is. Consider the analogy used by the philosopher Edward Feser: a metal detector is a perfect tool to locate metals, but that does not mean there is nothing more to this world than metals. That would be a silly conclusion. Those who protest that this analogy is no good, on the grounds that metal detectors detect only part of reality, while science detects the whole of it, are simply begging the question, for as to whether science really does describe the whole of reality is precisely what is at issue. An instrument can only detect what it is designed to detect. And that is exactly where scientism goes wrong: instead of letting reality determine which tools are appropriate for which parts of reality, scientism lets its favorite technique dictate what is considered "real" in life. Feser concludes from this, "That a method is especially useful for certain purposes simply does not entail that there are no other purposes worth pursuing nor other methods more suitable to those other purposes."

A sixth argument against scientism is that it is based on a very restricted view of reality. This can be explained by using an image borrowed from the late psychologist Abraham Maslow: If you only have a hammer, every problem begins to look like a nail. So instead of idolizing our "scientific hammer," we should acknowledge that not everything is a "nail." Even if we were to agree that the scientific method gives us better testable results than many other sources of knowledge, this would not entitle us to claim that the scientific method alone gives us genuine knowledge of reality. In other words, science can certainly do a lot for us, but its achievements are of a rather limited scope. As a matter of fact, science limits itself exclusively to what can be measured, dissected, counted, and quantified, and then declares anything else as nonexistent because it cannot be measured, dissected, counted, and quantified. Interestingly enough, the astonishing successes of science have not been gained by answering every kind of question, but precisely by refusing to do so. We shouldn't forget that science has purchased success at the cost of limiting its ambition—therefore, it can't claim *universal* validity for its *local* successes.

A seventh argument against scientism is that it turns out to be a form of circular reasoning. The late philosopher Ralph Barton Perry exposed its circularity as follows: "A certain type of method is accredited by its applicability to a certain type of fact; and this type of fact, in turn, is accredited by its lending itself to a certain type of method." That's how we keep circling around. After limiting itself exclusively to what can be measured, dissected, counted, and quantified, scientism then declares anything else as nonexistent because it cannot be measured, dissected, counted, and quantified. That is actually a form of circular reasoning. To use the analogy of a metal detector again: There is supposedly nothing more to this world than what a metal detector can detect. That's apparent nonsense. A thermometer, for example, can tell us what the temperature is, but it cannot tell us that temperature is all there is. As mentioned earlier, the Nobel Laureate Erwin Schrödinger once stressed it well, "It [science] knows nothing of beautiful and ugly, good or bad, God and eternity."

The eight reason to debunk scientism is its inconsistency. It declares that science is about material things, yet at the same time scientism itself is not something material. Scientism is one of those immaterial things that it denies they exist. First, scientists decide to limit themselves to what is material and can be dissected, counted, measured, and quantified. But then scientism kicks in and says that there is nothing else in this world than that which is material and can be dissected, measured, counted, or quantified. However, this verdict itself is not material and cannot be dissected, counted, measured, or quantified. It is again a kind of boomerang that comes back to hit whoever launched it. Nonetheless, scientific research requires immaterial things such as logic and mathematics.

Logic and mathematics are not physical, and therefore not testable by the natural sciences—and yet they cannot be ignored or denied by science. In fact, science heavily relies on logic and mathematics to interpret the data that scientific observation and experimentation provide. Yet, these immaterial things are real and indispensable, even though they are beyond scientific observation.

A ninth reason for rejecting scientism is that no science, not even physics, is able to declare itself a superior form of knowledge. Some scientists have argued, for example, that physics always has the last word in observation, for the observers themselves are physical. But why not say then that psychology always has the last word, because these observers are interesting psychological objects as well? As a matter of fact, neither statement makes sense; observers are neither physical nor psychological, but they can indeed be studied from a physical, biological, psychological, or statistical viewpoint— which is an entirely different matter. Even when physicists speak of a "grand unified theory (GUT)"—a theory that unifies the three non-gravitational forces—then this theory would definitely not be a theory of *all* there is in this world, but at best a theory of *physical* phenomena rather than a theory of *everything*. It's at best a theory about everything in physics.

A tenth problem of scientism is its underlying, yet often hidden, assumption that matter is all there is (whatever "matter" is). First of all, scientism itself is not a material entity. But more importantly, science is not possible without the use of two other non-material entities—logic and mathematics. G. K. Chesterton liked to ask his readers, "Why should not good logic be as misleading as bad logic? They are both movements in the brain of a bewildered ape?" If that were true, good logic or math would be as misleading as bad logic or math. However, logic and mathematics are not material and not physical, and therefore not testable by the empirical sciences. Yet, in logic and mathematics, immaterial things are true and demonstrable, even though they are beyond scientific observation. Hence, these immaterial truths cannot be ignored or denied by science, let alone scientism. Limiting oneself exclusively to material aspects of this world is in itself at best a metaphysical decision. One cannot give science the metaphysical power it does not possess.

A last argument against scientism—perhaps the strongest of them all—is that science only deals with causes that affect other causes. But the question remains: How are such cause-and-effect relationships possible? That's not a scientific question to be answered by using scientific tools. Responding that cause-and-effect relationships are ruled by scientific laws merely shifts the problem to the question as to where these laws come from. Scientism has nothing to say about this question—and yet cannot deny or discard that very question. The only reasonable response is what we discussed earlier (§5): secondary causes are only possible if there is a First Cause. This means there can be no science without the First Cause, because science is the study of secondary causes. This line of reasoning would be a final nail on the coffin of scientism.

In spite of all the above objections and arguments against scientism, it is an ideology still very much alive, albeit mostly unspoken or hidden underground. The late Dutch physicist Hendrik Casimir—the Casimir effect of quantum-mechanical attraction was named after him—once said, "We have made science our God." Indeed, science has become a semi-religion of which the scientists are considered the "priests." Science is supposed to explain everything, but then in a much better way than God could ever in their view. It is in this frame of mind that Stephen Hawking once exclaimed, "[O]ur goal is a complete understanding of the events around us and of our own existence." Scientism likes to broadcast to everyone around, "It's all about science." We cannot but ask, "Really?"

Well, science may be everywhere, but science is certainly not all there is. Interestingly enough, the astonishing successes of science have not been gained by answering any kind of question, but precisely by refusing to do so. Apparently, science has purchased success at the cost of limiting its ambition. As said earlier, it is true that if science does not go to its limits, it is a failure, but it is equally true that, as soon as science oversteps its limits, it becomes arrogant—a know-it-all, a form of gross megalomania.

Interestingly enough, when the *Royal Society of London* was founded in 1660, its members—the scientists of that time— explicitly demarcated their area of investigation and realized very clearly that they were going to leave many other domains of human interest untouched. In its charter, King Charles II assigned to the fellows of the Society the privilege of enjoying intelligence and knowledge, but with the following important stipulation: "provided in matters of things philosophical, mathematical, and mechanical." ("Philosophical" meant "scientific" back then; the term scientist was only coined in 1833 by the English philosopher and historian of science William Whewell.) That's how the domains of knowledge were separated; it was this "partition" that led to a division of labor between the sciences and other fields of human interest, including philosophy and theology.

By accepting this separation, science bought its own territory, but certainly at the expense of all-inclusiveness; the rest of the "estate" was reserved for others to manage. On the one hand, it gave to scientists all that could "methodically" be solved by dissecting, counting, and measuring. On the other hand, these scientists agreed to keep their hands off all other domains—education, legislation, justice, ethics, and certainly religion—for the simple reason that those require a different "expertise." Scientism, on the other hand, would deny and reject all of this, but without having any valid reasons to do so.

7. Science Is Never the Final Truth

Because scientific theories can never be proved but at best be disproved (§3), the process of determining "what the world is like and what it is not like" is inherently open-ended. It is science's aim to pursue this inherently open-ended process. As a result, science is a tentative enterprise, leading to acceptance, revision, or rejection of a hypothesis or a theory or a network of theories. Because the empirical cycle is a never-ending, cyclical process, science never ends. What is accepted today may be modified or even discarded tomorrow.

What then makes many of us think that science is *not* an open-ended process? Perhaps the main reason is that we learn about science through textbooks in school, which may give us the idea that this is the final state, and not merely the present state. However, *frontier* science is different from *textbook* science. Textbooks are able to select scientific information that has withstood the test of time so far, but journals and newspapers often contain new information that is less well established. Every reproduction of the present state of knowledge is just a picture at a given moment. This picture masks the fact that science produces merely temporary results in an ongoing process of developing more respected theories. The resolution of one question always generates more profound questions, so we found out (§3).

What obscures this important point even more is the fact that when publishing findings in a scientific journal, scientists tell us only the end results of their research, but seldom or never all the blind alleys they had to go through. Yet, publishing scientific results is pivotal in science. The physicist Michael Faraday told his colleagues: "Work, Finish, Publish." Scientific research is not complete until its results have been published. But publication is never the end of the story.

But what is even more serious is that we often only see the positive results published, while missing the negative ones. It has been found, for instance, that statistically significant results are three times more likely to be published in the medical field than papers with results that are not statistically significant. Results not supporting the hypotheses of researchers often go no further than the researcher's file drawer. Although it's definitely true that science can do a lot for us, the process doesn't go as smoothly as often portrayed. It's rather a process of trial and error that's jerking along.

Yes, experiments can fail immensely, interpretations of experiments can be flawed, and scientists can make mistakes. Yet, the nature of science is self-correcting. Scientists keep devising theory after theory. What drives them to do so? Albert Einstein probably gave us the best answer: "Why do we devise theories at all? The answer to the latter question is simply: because we enjoy 'comprehending,' i.e., reducing phenomena by the process of logic to something already known or (apparently) evident." In his 1939 lecture, *The Usefulness of Useless Knowledge*, Abraham Flexner, best known for his role in the 20th century reform of medical and higher education in the United States and Canada, stressed the fact that great scientific discoveries were not necessarily made for their usefulness. They were rather the product of some kind of "curiosity" that sought no rewards (other than the Nobel Prize and fame).

In other words, science is a perpetual process of finding and then searching again, mostly driven by an intellectual desire for knowledge—in spite of the fact that we have been bombarded with claims concerning the upcoming "completion" of science and the imminent "unification" of theories. In spite of the fact that science is always a work in progress, some scientists cannot resist the temptation to claim certainty and finality. The Dutch physicist Pieter Zeeman, later to become a Nobel laureate, was fond of telling how, in 1883, when he had to choose what to study, people had strongly dissuaded him from studying physics. "That subject's finished," he was told, "there's no more to discover."

It is even more ironic that this also happened to Max Planck, since it was he who, in 1900, laid the foundations for one of the greatest leaps in physics, the quantum revolution. He was told during his training that physics had been completed and finished. In fact, what was perhaps "finished" is what we now call classical physics. It is to be doubted, though, that it is in the nature of science ever to arrive at completion. Yet it remains a timeless temptation to claim that the unknown has been reduced to almost nothing. However, the magnitude of the unknown is, well... unknown! As the late physicist John A. Wheeler put it, "As our island of knowledge grows, so does the shore of our ignorance."

As a consequence of this open-endedness of science, we must also accept the possibility that certain "scientific facts" may turn out not to be facts at all (§8). That is the reason why many current "scientific statements" may have to face the prospect of becoming outdated. It is a frequent saying that certain statements in widely used textbooks are not quite true, or no longer true at all; that certain sections have become outdated, etc. That's how science grows. A snapshot picture of science only show us a part of the much larger process at a certain point in time.

In general, science grows in fits and starts. Sometimes the

changes are rather minor. In fact, there are many scientific theories that were believed to be true, but then sooner or later turned out to be false—they were "facts" that turned out *not* to be facts. Here is just a small selection of such cases. (1) Vulcan was a planet that nineteenth-century scientists believed to exist somewhere between Mercury and the Sun in order to explain certain peculiarities about Mercury's orbit (case #9). (2) The expanding Earth hypothesis stated that phenomena like underwater mountain ranges and continental drift could be explained by the fact that the planet was gradually growing larger. (3) Prior to scientists embracing the notion that the Universe had begun with the Big Bang, it was commonly believed that the size of the Universe was an unchanging constant. The Static Universe is also known as "Einstein's Universe," because he argued in favor of it and even calculated it into his theory of general relativity. (4) As peptic ulcers became more common in the twentieth century, doctors increasingly linked them to the stress of modern life, until the Australian clinical researcher Barry Marshall discovered that peptic ulcers are caused by the bacterium *Helicobacter pylori*. (5) For some thirty years, the number of human chromosomes was supposed to be 48, until the geneticist Joe Tjio found it to be 46 in 1955. (6) Until the mid-twentieth century, most paleoanthropologists preferred Asia, over Africa, as the continent where the first hominids evolved. The recent African origin of modern humans ("Out of Africa") is the currently preferred theory.

But there are also times when science goes through quite dramatic changes—so revolutionary that it may take a while for them to be accepted by the scientific community. The history of science shows us that such drastic revisions may not be accepted until the defenders of an outdated outlook—the representatives of an "old school," or an "exploded paradigm"—

have dropped out of the picture once and for all. Here are some examples. William Harvey never saw his theory of closed blood circulation fully accepted during his lifetime. Isaac Newton's theory that gravitation is responsible for the motion of the planets required some eighty years before it was universally accepted. Alfred Wegener's theory of continental drift was published in 1912, but was generally adopted only fifty years later, after the acceptance of the theory of plate tectonics. For almost thirty five years, Gregor Mendel's findings escaped notice until geneticists were able to appreciate them (case #4). It took decades for antiseptics—discovered by Ignaz Semmelweis in 1847 (case #5), by Louis Pasteur in 1862 (case #6), and by Joseph Lister in 1867—to become generally accepted in surgery. And then there is Einstein's 1905 work on relativity which remained controversial for many years, until it was adopted by leading physicists, beginning with Max Planck. And the list could go on and on.

Another dream of scientists, especially of the physicists among them, is establishing one Grand Theory unifying all physical theories—the so-called *Grand Unified Theory* (GUT). These scientists are in search of a physical "umbrella" theory that would unify the three non-gravitational forces—namely the electromagnetic force, the strong nuclear force, and the weak force—as aspects of one single underlying force. It is merely a dream that is not yet reality, in spite of many trials made on the "drawing board" to combine Einsteinian gravity and quantum theory. Probably the best of these trials is *superstring* theory. We do not know whether superstring theory is the true theory, and we may never know, but many physicists believe they have good reasons to assume it is.

Can science ever accomplish this goal of finding a "grand unified theory"? Only time can tell; it's an empirical question. But as we said earlier, even if or when science achieves this goal, its would-be "grand unified theory" is definitely not a theory of *all* there is in this world. It would at best be a theory of *physical* phenomena, not a theory of *everything*. There is no science of "all there is." A theory of everything would also have to explain why some people believe it and some do not. One cannot give science, nor any of its theories, the metaphysical power it does not possess. Physics, for example, discloses only one of the many aspects of material world, its physical aspect, thus providing physical answers to physical questions phrased in terms of physical causes and effects. Physics may be everywhere, but it's surely not all there is. Biology offers another perspective, providing biological answers to biological questions in terms of physical causes and biological functions. But there is no science of "all there is" (§6).

All these considerations combined make it unlikely that science can ever reach the final truth. Another reason why science can never get to the "final truth" about this world is the fact that it is "blind" for many aspects of life that are non-material. Science may have its own window on the world, its own scientific point of view, its own perspective, but there are many other windows, views, vistas, perspectives, or whatever you wish to call them.

Reality is like a jewel with many facets that can be looked at from various angles and with different eyes. In a similar way, there are many perspectives, aspects, and outlooks on the world surrounding us, making for different points of view, different conceptual frameworks, different angles, and different facts— all of which can be equally "real," "factual," "objective," and "valid." Yet each one offers us only a one-sided perspective on the multi-faceted reality we live in (§14). They are all biased on some kind of conceptual frameworks. These frameworks are networks of theoretical beliefs about what kinds of entities exist.

Because science offers us only a scientific perspective on the world, there is so much more to life that is off-limits for science. There are many more perspectives than what science tries to cover with its telescopes, microscopes, barometers, thermometers, and spectrometers. We shouldn't close our eyes to what science has left untouched. We may neglect what lies beyond science's horizon, but we can't therefore validly reject or deny what we neglect, as is done in scientism (§6).

So reality can be looked at from various angles, from different viewpoints. Just like the "physical eye" sees colors in nature, so the "artistic eye" sees beauty in nature, the "rational eye" sees truths and untruths, the "moral eye" sees rights and wrongs, the "philosophical eye" sees everything in relationship to the First Cause, and the "religious eye" sees everything in relationship to God. All these "eyes" claim to be in search of reality, but each one "sees" a different aspect of it—and therefore sees different facts (§8).

What can we learn from this for the education of our new generation? Students in our schools—from elementary school to college—deserve to be taught genuine science, so they and their parents should not settle for some kind of ideology. Hence, in teaching science, it should also be made clear what its limitations are—such is part of teaching genuine science as well. The motto for teaching science should be: teach it, but do not preach it, otherwise science becomes a semi-religion. Science becomes involved in an identity crisis, says the physicist Fr. Stanley Jaki, "when it ignores its own method or when it lets philosophers, eager to promote their agnosticism and subjectivism, take over as the spokesmen for science." That would be a real and serious disservice to science.

III

How to Get the Facts?

Science could be seen as a *fact*-finding endeavor. However, getting to the facts is not as easy as opening your eyes. That is the reason why there is so much discussion about "the facts" in science, as we found out already (§4).

Discussing the facts is very common in science. So it should not really surprise us that the high-standard scientific journal *Nature* recently published a disturbing commentary claiming that in the area of preclinical research—which involves experiments done on rodents or cells in petri dishes, with the goal of identifying possible targets for new treatments in patients—independent researchers doing the same experiment could not get the same results as reported in the scientific literature. Over a period of ten years, for example, Amgen researchers could reproduce the results from only 6 out of 53 land-mark papers. And researchers at Bayer Healthcare reported that only in 20-25% of 67 projects analyzed the relevant published data were completely in line with their in-house findings.

So what are we to say then about the scientific facts we hear about and read about? At first sight, they do not seem to be as solid as portrayed by scientists. How come? I have to warn you, to answer that question we need to delve into some philosophy again, and you will soon find out why.

8. What Are Facts?

It is through scientific research that we found out the many facts science has given us—the facts of science, the scientific facts. Examples are: "The Universe started with the Big Bang," "Stars are factories of chemical elements," "Planets orbit stars in elliptical paths," "The atomic nucleus contains 99.9 percent of the mass of the entire atom," "Photosynthesis releases oxygen into the atmosphere," "Human cells contain 23 pairs of chromosomes." You can find these scientific facts in textbooks, encyclopedias, journals, and on the internet.

Science is known and praised for giving us "scientific facts." That which makes them scientific is the empirical cycle. But what is it that makes them *facts*? We discussed already (§7) that scientific facts may turn out, on further inspection, *not* to be facts. Just like there are simple ideas about science (§1), so there are also simple ideas about facts.

Most people tend to think and talk about facts in a very casual way, overlooking the very "fact" that facts are very peculiar entities. Facts are not things we can touch or smell or even see. They are not, as so many think, the rock-solid pieces of this world that we "bump" into. They are not just "there" waiting for us to open our eyes—if so, that would have made science so much easier. So what then are they?

The first problem is that what is "out there" is not an assortment of rock-solid facts, but rather a collection of things, situations, processes, and events. Facts, on the other hand, are not any of these; they are rather our interpretations of the things, situations, and events we encounter around us—in short, facts are our means of making them *intelligible* for us. There are simple facts such as "Snow is white," or more intricate ones such as "Litmus paper turns red in acids," or rather sophisticated ones such as "Genes are made of DNA," or complicated ones such as "Radioactive isotopes decay with a constant half-life rate."

Things, situations, processes, and events may be the "material" parts of our world, perhaps even rock-solid, but facts are "non-material" entities. Let's analyze this vague distinction further by contrasting facts respectively to events (1), thoughts (2), and statements (3). I apologize for this highbrow philosophical analysis, but we need it for what is coming.

(1) Facts are different from *events*. Unlike facts, events are dated, tied to space and time, whereas facts are time-less and space-less. The fact that planets follow elliptical orbits, for instance, is not dependent on a certain time or space. It is even considered a fact that certain events did not occur or that certain things do not exist. It is a fact, for instance, that Darwin and Mendel never met, but there is no event at a certain time that we can point to. It is also a fact that centaurs do not exist. (2) Facts are also different from *thoughts*. Thoughts can be imaginary, illogical, confused, time-consuming, and so on— whereas facts cannot be any of these. Facts are true, even if some people have never thought of them, or even if no one has ever thought of them. Facts are true even when they have not been discovered yet. (3) Facts are also different from *statements*. Statements can be hypothetical, inaccurate, exaggerated, long-winded, and difficult to understand, whereas facts cannot be any of these; a fact may be hard to accept, but never hard to understand; a fact is true, but never hypothetical or half-true. The statement about a fact may be hard to understand, but the fact itself is not. There are even facts which everyone has forgotten or which were never expressed in a

statement yet. Facts are beyond our control, but statements we can make on our own at any time. So we must come to the following conclusion: a fact is not an event, not a thought, and not a statement.

We are facing here a rather complex situation: if facts are not events nor thoughts nor statements, what then are they? Facts actually feature as a focus point at the intersection of those three other elements: a fact is the description of an event, the object of a thought, and the content of a statement, all at once. Facts are closely connected to these three other elements through the process of interpretation by the human intellect: facts are interpretations of events by means of thoughts and statements. It is through interpretation that thoughts and statements transform events into facts. However, facts may need events so they can be tested; they may need thoughts so they can be understood; and they may need statements so they can be communicated. But the bottom-line is this: unlike things, events, or situations, facts are non-material entities.

It should not surprise us then that a camera, for example, cannot capture facts—all it can capture is things, situations, and events. Take a surveillance camera; it "observes" everything because it does not "know" what to observe. That is why cameras and other observational tools cannot replace scientists—they may be helpful to them but cannot replace them. The problem with pictures in general is that they do not show us facts until we give some interpretation to what is "seen" on a picture. The same with books: they can provide lots of factual information for "bookworms," but to real worms they have only paper to offer.

In other words, facts carry a heavy man-made component—it's called interpretation. This does not mean, though, that facts can change—facts are facts, period. But sometimes we declare something a fact, which turns out on further investigation *not* to be a fact. The same point can be made in terms of knowledge. Like facts, knowledge is "infallible." Whereas beliefs can be either true or false, knowledge can only be true if it's real knowledge. If I believe something to be a fact, which it is not, I don't conclude I had false knowledge, but that I did not know at all.

We found out in the first chapters that the indubitable foundation of knowledge, scientific or otherwise, cannot be found in observation statements. When speaking of facts, we claim much more than what we observe, because there is no factual information without interpretation by the human intellect—and as a consequence, a lack of "interpretation" amounts to a lack of "information." The more interpretation we inject, the more information we must provide, but also the more we need to confirm. For example, when we describe a certain event taking place in the sky as "Those are moving spots," we do express a fact, but it contains pretty "empty" information, which makes it a rather "safe" form of information. When we say, however, "Those are flying birds," our statement does convey additional information—and therefore we may need to come up with more evidence to support our claim. And when we say, "Those are migrating geese," we inflate our information even further. We have three different facts here regarding the same situation.

In other words, in defiance of inductivism (§1), scientific observations are not like pictures taken by the "camera of our eyes," for even the pictures taken by a real camera would have to be interpreted first before they could convey any information in the form of facts. Science, for instance, is one of our main tools to make the world understandable and intelligible with the help of scientific concepts, hypotheses, and theories (§3). Although science is exclusively about the material world, the world of scientific facts, laws, and theories is an non-material world—dealing with and referring to the material world, but definitely distinct from the material world. Facts require concepts, and concepts require a human intellect. Even simple facts do. The fact that snow is white, for example, requires the concepts of snow and whiteness, and those concepts come from the human intellect.

One word of caution, though. The term "interpretation" may put some people on the wrong foot by making them think we are projecting something into the world that is not there. "Interpretation" does not mean that facts only exist inside our minds—far from that. Of course, a specific interpretation made by the human intellect may be wrong; it may reveal "facts" that are not facts. That's why our interpretations—concepts, hypotheses, and theories—need to be tested so as to find out whether they revealed facts rather than illusions or wishful thinking. Remember, not all inventions lead to discoveries.

In other words, what we call proven scientific knowledge is only proven until a new set of empirical facts disprove what was previously considered proven—which we called falsification before. When Francis Crick said, "A theory that fits all the facts is bound to be wrong, as some of the facts will be wrong," he was right, but not precise enough in what he said. To express this idea more accurately, we should say that facts cannot be wrong or false, but they may turn out *not* to be facts. It is always important to know what is not true and why it is not true, to know what is knowledge and what is not knowledge.

Some have objected, though, that the facts we know are part of our knowledge and therefore cannot be external to our knowledge—which leads them to think that speaking in terms of anything "outside" our knowledge would be out of the question. In such a view, however, knowledge is conceived of as if it were a box, allowing things only to be either inside or outside the box. On the contrary, knowledge is not like a box. It may be best compared to a source of light, as the famous philosopher Edmund Husserl saw it: if a light beam hits a certain thing that is in darkness, this thing will be in the light, and yet it won't be inside the source of light. In translation, research is first of all a matter of asking the right questions—by using the searchlights of concepts, hypotheses, and theories. The best way to search is to have an idea of what you are looking for. Searching thrives on ideas; without preliminary ideas, science would be blind. But ultimately, test results determine whether those ideas do reveal scientific facts.

Seen this way, scientific concepts, hypotheses, and theories could be considered to function as "searchlights" which may bring to light facts that have escaped notice so far because the human intellect had the wrong interpretation before. They may help us find what we were searching for—but not necessarily so, for searching does not always end up in finding. Earlier (§2), we mentioned Plato's paradox: We are in search of something "unknown"—otherwise we wouldn't need to search anymore—and yet it must be "known" at the same time—otherwise we wouldn't know what to search for, or wouldn't even know if we had found what we were searching for.

That is the reason why we need provisional, hypothetical statements based on concepts, hypotheses, and theories which work together like "searchlights" to help us illuminate what was in "darkness" for us before. That's how facts come about, in an interaction between thoughts, statements, and events. Although all scientific searchlights are immaterial, what they search for is material.

Because each scientific field has its own interpretations, it also deals with its own facts. A psychologist has an "eye" for psychological facts, whereas a biologist perceives only biological facts. This does not mean that our world is compartmentalized, but science is. The very same event can be looked at from different "angles" or "perspectives," with different "glasses," within different frames of reference—or whatever image you want to use. In a sense, science makes for a fragmented world. Nevertheless, all sciences deal with the same world, external to us and to our minds. That world is the final touchstone of all our interpretations. Ultimately, "reality" is the final foundation of all we know; so it should always be the ultimate touchstone of all our interpretations.

As to be expected, we don't find facts only in science. First of all, there are facts in daily life that we don't need science for, such as the fact that snow is white. Second, there are also facts in religion that science has no access to: facts such as "God exists," "God created the universe," "There is life after death," "Humans have a soul," "Humans are immortal," "Jesus is the Son of God," "There is a Last Judgment." These are facts that don't come from science, but instead from religion, often more in particular from Scripture. Whether it's in science or in religion, we are dealing with facts that are beyond our control—if we deny them, they don't cease to be facts, and if we don't know them, they don't stop being facts. Stating, for instance, that the earth is flat doesn't change the fact that the earth is *not* flat. Stating that God does *not* exist doesn't change the fact that God does exist.

That facts do exist in religion may surprise many people. Yet, religious facts share with scientific facts that they are interpretations of reality. They are actual and true—they are both about reality. If they are not true or real, then they are not facts. We may have thought for a while that they were facts but they turned out *not* to be facts. In other words, there is no opposition between science and religion in the sense that one of them, science, deals with facts, whereas the other does not.

Put differently, religious faith is not just a matter of a personal belief or a subjective opinion—it is as much about objective facts as science is about objective facts. The Nicene Creed, for instance, is actually a litany of facts—e.g. about God as "maker of Heaven and Earth, of all things visible and invisible," about Jesus who "will come to judge the living and the dead" and who "is seated at the right hand of the Father," and about "one Baptism for the forgiveness of sins." Granted, religion is more than a matter of facts, it certainly is not less than that.

In other words, why should we believe something, even in religion? Not because others believe it. Not because it feels good. But only because it is true—because it's a fact! You may have your own personal beliefs, but you cannot have your own personal facts. The existence of God is a factual matter for everyone: either God exists or he does not. Just as believing that the earth is flat does not make the earth flat, so believing that God does not exist does not make God disappear. If it turns out God doesn't exist, it isn't that our faith was wrong—our facts were wrong, they were not facts.

Where science and religion do differ, though, is how we know those facts. In science, we test them through the empirical cycle, because science deals with material entities (§3). But in religion, we cannot test that way, because religion is about invisible, non-material entities. We cannot test them in the laboratory or through experiments—that's something we do for secondary causes, but certainly not for the First Cause. The question "Does God exist?" is not like "Do neutrinos exist?" God, being the First Cause, cannot be "trapped" by some kind of ingenious experiment. Yet, we still can, and must, test religious facts, but that's done with logic and reason, among other tools.

Reason provides the "litmus test" for religious facts. The famous proofs of God's existence, for instance, use logic and reason to show that God's existence is not only very reasonable but actually extremely compelling (§13). It is not a scientific proof—if there is such a thing—but rather a metaphysical proof that comes with "certainty beyond a reasonable doubt." As we said earlier, it provides deductively conclusive support, whereas science can only provide inductive evidence.

St. Thomas Aquinas went even as far as proving that no religious faith, even if it comes from the Bible, should ever be *against* reason and logic. If something is against reason, it cannot be from God. God cannot do what is contradictory, because God cannot act against his own nature, which includes Divine Reason. Even the so-called mysteries of the Catholic Faith may go beyond mere reason, but they are never against reason—they are not even unreasonable. A mystery is not something about which we can't know anything, but something about which we can't know everything. The reason that there are mysteries is that God is infinite but our intellects are finite.

People who swear by the power of experiments typically tend to underestimate the power of reason. But sometimes the power of reason can help us prove or disprove something without needing any experiment. Let me show this with the following case that I borrowed from Peter Medawar. Anyone looking at paintings of the famous Renaissance painter El Greco will notice that most of his figures are unnaturally tall and thin. Some scientists came up with the hypothesis that El Greco must have had a form of astigmatism that distorted his vision and led to elongated images forming on his retina. Others tried something similar to explain Vincent Van Gogh's preference for yellow colors by hypothesizing a visual disorder called xanthopsia; some even mentioned drug use or glaucoma as the causing agent.

In cases like these, the mere use of reason could have saved these scientists embarrassment. Even if El Greco did see the world through a distorting lens, the same distortion would apply to what he saw on his canvas. These two distortions would cancel each other out, and the proportions in pictures would remain realistic. Even if Van Gogh did view the world through a yellow filter, he would also view the colors on his canvas that way. In either case, reason and logic could have proven ahead of time that the scientific explanations mentioned above are doomed to fail, even before any scientific test had been done.

What we can learn from this is that reason, logic, and philosophy are important tools to learn about the facts. Not only can they disprove what we thought were facts, but they can also prove certain facts. The best example of the latter are the proofs of God's existence. Let me just use one simple example— but there are many more. It runs as follows: nothing that came to exist can cause its own existence, for to do so it would have to exist before it came into existence—which is logically impossible. Therefore, whatever came into existence must have a cause outside itself that caused its existence. This is basically a deductive form of reasoning—sometimes called *reductio ad absurdum*. In this sort of argument, we begin by assuming the opposite of the claim we want to prove, and show that from this claim something absurd follows. This shows that the opposite of the claim that we want to argue *for* is false, and so that the claim we want to argue *against* is true.

From this starting point, we can derive much more. If nothing can be the cause of itself and cannot impart existence to itself, then its existence must be imparted by some cause distinct from it. Even if we accept that whatever came into existence must have a cause outside itself that caused its existence, then we still have not explained how those causes can be causes of their own. So there must be a First Cause that causes them to be causes of their own. This constitutes a conclusive rational argument for the existence of an infinite and necessary First Cause, which is un-caused, because it is eternal, never came into existence, and therefore is not in need of any other cause.

Many other proofs of God's existence have a similar approach. Let us discuss one more argument, the "argument from contingency," because the four other proofs that St. Thomas mentions in his Five Ways could be taken as variations of this one particular proof. The point of departure is that our Universe need not be the way it is, and need not even exist. The Universe contains *contingent* beings who could easily not have existed, as the reason for their existence cannot be found within themselves—they cannot make themselves exist. Therefore, they depend for their existence on an overarching, transcending "ground"—a Necessary Being. Existence is something received. However, if there is no inherent necessity for the Universe to exist, then the Universe, including everything in it, is not self-explanatory and therefore must find an explanation outside itself. Obviously, it cannot be grounded in something else that is also finite, contingent, and not self-explanatory—for that would lead to infinite regress—so it can only derive from an unconditioned, infinite, and necessary ground, which is a "Necessary Being," not self-caused but un-caused.

The existence of God is just one of the facts religion deals with. How do we know there are other religious facts? The answer goes like this: partly through reason, partly through personal experience, partly through God's Revelation, partly through Scripture, partly through the Magisterium of the Church. All religious facts have to go through these tests to make sure they are really facts and not illusions. In other words, religious beliefs are not just based on ideas, but on facts. Either there is a God or there is not. Either there are souls or there are not. Either there is life after death or there is not. Either there is a Last Judgment or there is not. Either Jesus is the Son of

God or he is not.

The apostle Paul spoke in the same vein when he said "[I] testified of God that he raised Christ, whom he did not raise if it is true that the dead are not raised. For if the dead are not raised, then Christ has not been raised. If Christ has not been raised, your faith is futile" (1 Cor. 15:15-17). There is no faith without facts.

Just like all other facts, so are religious facts either true or they are false. They are either facts or they are not facts. However, what sets religion apart is that it sees everything in a religious perspective—that is, in relationship to the Creator and First Cause of this world, whereas science see the world in relationship to secondary causes. But that difference doesn't make a religious fact less of a fact.

Unlike animals, which only live in a world of things and events, human beings also live in a world of facts. By stating facts, humans always claim much more than what they "observe." In observation, one is both a passive "spectator" and an active "creator" at the same time—no matter whether it is in science or in religion. In either case we are dealing with facts. There could be no science without them and there could be no religion without them.

9. From Concepts to Facts and Back

Facts are usually expressed in statements. Whereas a statement can be wrong, a fact cannot be wrong—it just is not a fact. Statements are possible thanks to the concepts they use. It's due to concepts that we can see what cannot be seen without them. And yet, concepts cannot be seen themselves—they are abstract and immaterial. You can see round things, pick them up, put them somewhere, but you cannot do any of these with the *concept* of roundness. Although it is an object of thought, it is not a concrete object like a red ball, but instead it is an abstract object.

Concepts are the building blocks of facts. We can't think about facts without using concepts, and we can't talk about facts without using words in statements. Most words designate one or more specific concepts; or reversely, concepts are the meaning "behind" words. Words are merely "labels." But labels for what, you might ask. The shortest answer is: words are labels for concepts. The same concept may be connected with different words in different languages. For instance, the word "Snow" in English and the word "Schnee" in German are two different labels for the same concept—snow. The underlying concept for these two labels is the same, no matter which language we use.

But this immediately raises the question: What are concepts then? A mistaken view of concepts is that they are "pictures" in our minds. But what would a "mental picture" behind the concept of "dog" look like? Would it be a composite picture of an ideal terrier plus an ideal collie plus of all the other breeds? Earlier we discussed a safer alternative: concepts are like miniature theories (§3). Instead of saying that concepts emerge from observations, it is much more likely that observations are based on concepts and are only possible because of concepts.

Concepts are our main tools to identify and categorize things in the world around us. By doing so, they allow us to see similarities between things that would otherwise elude us. Facts may be staring us in the face but we do not "see" them if we don't have the right concepts yet. This is true of simple concepts such as "snow" or "white" as well as complex scientific concepts such as "proton" or "gene."

Let's start with rather simple concepts. As shown earlier, in order to make any observations, we depend on concepts. When we see "snow," for example, we classify it in the category of things that are white, that are cold, that can melt, etc. Thus, the word "snow" is part of a larger conceptual network which makes for a miniature theory. Once a concept has been acquired, one may easily forget that its meaning and usage had to be mastered. Without knowing it, one has become a "regular observer" of snow, for instance. It is through concepts that perceptual experiences become conceptual observations. Concepts are meant to "cut out facts" in the world we live in; as concepts change, the world is gradually cut out in a different way, because concepts determine which things will be combined, and which will be distinguished and separated.

Without concepts we listen but do not hear, we look but do not see, we are blind with our eyes open. A concept is a kind of preconception of what some part of the world looks like. The act of generalizing from a few observations to all similar observations is based on the fact that objects look alike in certain respects—for instance, being white—but the similarity of white things is not visible until we have a concept of being white. We cannot simply master this concept by pointing at something white, for mere pointing can identify many things but not necessarily the "whiteness" part.

Concepts are always dispositional and lawful; they classify by putting something in a category of similar things; they discriminate and set something apart. Because concepts involve regularity and similarity, they cannot possibly hold for only one single case. Even the concept of "uniqueness" does not apply to one single case, but puts a thing into the category of things that are unique in one or more respects.

What we said about simple concepts basically also holds for scientific concepts. As we discussed earlier, science does not start with observation, but instead it is the other way around: observation starts with concepts—it is "concept-laden."

Because concepts have a more or less intricate relationship to other concepts, they can be considered low-level or miniature theories. Hence, we should rephrase our previous conclusion in this way: observation starts with theories, rudimentary as they may be. It should be clear by now that the transition from concept to theory is a smooth one. A concept hides a miniature theory, and a theory in turn is a more elaborate system of concepts. In short, all observation is not only "concept-laden" but also "theory-laden." The astrophysicist Arthur Eddington put it this way, "Do not put too much confidence in experimental results until they have been confirmed by theory." Without concepts and theories, experiments don't make sense.

The history of science is basically a process of finding better concepts to describe scientific facts. A classic example was the chemical issue of combustion in the debate between Joseph Priestley (1733-1804) and Antione Lavoisier (1743-1799). Prior to Lavoisier, phlogiston was a standard concept in the theory of combustion. The old theory said that phlogiston is emitted from substances when they are being burnt. However, it was soon discovered that many substances actually gain weight after combustion, which seemed to be a falsification of the theory. Was this the end of the theory based on the concept of phlogiston? No, falsification was averted by the "ad-hoc" statement that phlogiston has negative weight, similar to the way the Ptolemaic system used to be saved by introducing epicycles.

Finally, however, the theory was falsified when Lavoisier introduced a new concept, the concept of oxygen. Seen in the light of the new concept of oxygen, the old concept of phlogiston faded away. While Priestley had seen phlogiston transiting from the fuel to the air, Lavoisier saw oxygen moving in the opposite direction, namely from the air into the fuel. The new concept could explain, for instance, why a candle under an inverted cup stops burning when the oxygen has been used up. Nowadays any chemistry student who is electrolyzing water is not manipulating phlogiston, but producing oxygen at one electrode and hydrogen at another. This is not just a matter of different words but of different concepts. Oxygen is more than the mirror image of phlogiston—in fact, a conceptual revolution has taken place.

The history of science is full of similar examples where conceptual revolutions propelled science and uncovered new facts. Biologists, for instance, were not able to see the similarity in building blocks between animals and plants until the concept of "cell" had been established. In the beginning the cell concept had hardly any content; what Robert Hooke saw was more of a dead or lifeless cell-wall than a cell. Through the work of Matthias Schleiden and Theodor Schwann, the cell became a structural unit (1839); and since Rudolf Virchow it is also a functional unit (1855). It was Virchow who interpreted every organism as a cell state, composed of physiologically interdependent cell units.

But eventually, Virchow's anatomical building-block cell theory had to yield to the concept of cell as a biological unity. The cell had become an organism in itself. Ever since, the cell has been viewed as the seat of the morphological and physiological unity of the body of animals and plants—actually the very unit of life. As frequently occurs, a great conceptual battle won is often a new scientific phase begun. The theory of cells has become a central, basic science—cell biology or cytology—from which many other areas of research were derived since.

Another case in the history of science is the issue of fermentation. Louis Pasteur considered all fermentation to be of a biological nature (case #6). Although he acknowledged that digestive ferments can function outside the cell—in an "unorganized" way, so to speak—he distinguished a different kind of fermentation, carried out by "organized" ferments such as yeast. It was Isidor Traube (1860-1943) who was able to expand the concept of ferment. He came up with the idea that "organized" ferments are nothing but "unorganized" ones, but functioning within the organized setting of a cell. The chemist J.J. Berzelius (1779-1848) had made the same move earlier by taking both kinds of ferments together under the heading of a new concept, "catalysis," soon to be followed by the concept "enzyme."

We saw something similar happen when the gene concept went through quite a conceptual change (§4). Starting as an element (1865, Mendel), it became a factor (1900, Bateson), then a unit of transmission (1909, Johannsen), as part of a chromosome (1915, Morgan), with a specific order of DNA bases (1953, Watson and Crick), contributing to the production of enzymes (1940, Beadle and Tatum), or of proteins in general (1957, Benzer), thus having a structural or regulatory function (1961, Jacob and Monod), and consisting of exons and introns (1977, Chambon). As a result, the gene concept has moved far away from what Mendel had envisioned.

What these examples—and there are many, many more—have in common is that new concepts can help scientists see similarities they could not see before. They could not see the similarity between leprosy and tuberculosis until the concept of bacteria had become available. They could not see the similarity between falling apples and revolving planers until Isaac Newton introduced the concept of gravity. Concepts act like "searchlights," as Karl Popper used to call them. They start as "inventions" which may have to be abandoned, but they may sometimes lead to "discoveries." Only research can tell us which searchlights reveal facts. They may help us find what we are searching for—but not necessarily so.

How to Get the Facts?

Obviously, when research does reveal new facts, those facts already existed before the new searchlights brought them to light, but we cannot say much about what is still in "darkness" for our intellect. Over and over again, we see concepts in science start as inventions. When the physicist Isidor Rabi was told about a new particle, the muon—an elementary particle similar to the electron, not thought to be composed of any simpler particles—he replied, "Who ordered that?" It turned out to be one of those concepts that went from an invention to a discovery. And thus, new facts were seen.

Concepts help us to see facts which we could not see before we had those concepts. This doesn't mean concepts create these facts, for they were already there before they were discovered, but they couldn't be "seen" yet. It may sound strange, but concepts such as gravity, atoms, isotopes, electrons, muons, cells, bacteria, genes, enzymes, and the like, existed already before these concepts became available. That's what science can do for us: discovering new concepts, and hopefully new facts, that had been eluding us, until we came to know about them.

Something similar can be said about religion. The concept of "original sin," for instance, refers to one of the pivotal facts of religion in Christianity. It can only be understood in the light of the salvation Jesus "bought" for us through his crucifixion. G. K. Chesterton called it "the only part of Christian theology which can really be proved." It is strongly connected with other concepts such as incarnation, crucifixion, suffering, sacrifice, salvation, and redemption. If there is no original sin, then the Cross of Golgotha is a hoax; if the Cross is a hoax, then the whole economy of salvation is up for grabs. In the words of the *Catechism* (389): "The doctrine of original sin is, so to speak, the 'reverse side' of the Good News that Jesus is the Savior of all men, that all need salvation and that salvation is offered to all through Christ." The first Christians had to develop these concepts before they could interpret Jesus' crucifixion and "see" what the event of the crucifixion really meant.

It is most likely that many more concepts and facts will be unearthed when science and religion further enhance our knowledge. But the extent of how much is still unknown, hiding in darkness, remains unknown until we have acquired the right concepts to search with. Science is in a permanent pursuit of searching in the hope of finding—and so is religion in its own way, at least in the doctrinal interpretations of the history of God's salvation, guided by the apostles, early Church Fathers, popes, Church councils, and theologians. They are both in search of facts.

10. Can the Facts Change?

When Francis Crick said, "A theory that fits all the facts is bound to be wrong, as some of the facts will be wrong," did he really mean that facts are wrong and can change? If so, that would be a violation of the very fact that facts never change. A fact is a fact—whether we like it or not, whether we believe it or not, whether we accept it or not. A fact is a fact, even before we knew it. It has always been a fact that stones fall to the ground, even before Newton told us why. It has always been a fact that human beings have cells with DNA in it, even before Watson and Crick told us so. No, it's not the facts that change; they are facts and remain facts—they don't and can't change being a fact.

What may and does change, however, is that some facts we thought were facts turned out *not* to be facts, and facts we didn't yet know about turned out to be facts. For Priestley, it was a fact that phlogiston is emitted from substances when they are burnt, until Lavoisier showed that this "fact" was not a fact. For

Galileo, it was a fact that the water in a water tower is "sucked up" by a "horror vacui" (nature's abhorrence of a vacuum), until Torricelli showed that it is "pushed up" by atmospheric pressure instead. Before Einstein, it was considered a fact that light always propagates in a straight line, but that's no longer a fact under all circumstances.

The viewpoint that facts are about reality and that facts cannot change is strongly connected with the assumption of *realism* (§14), which claims that science aims at true descriptions of what the world really is like. Realism claims that the world exists independently of us as knowers and of our theoretical knowledge of it. So-called metaphysical realists believe that the world consists of mind-independent objects. The goal of science is to discover and name these objects. When the light of a star reaches our eyes, it may have traveled a very long distance, yet we know the star was already there before its light reached our eyes. Who would ever deny the real existence of stars, electrons, molecules, cells, genes, and the like? In this view, a theory can be true even if nobody believes it and can be false even though everybody believes it. Luckily, we have minds that are open to what is not ourselves. We are called to understand as true what is not ourselves but something outside of us.

In his discussion with fellow physicist and friend Niels Bohr, one of the founding fathers of quantum mechanics, Albert Einstein is reported to have asked Bohr whether he realistically believed that "the moon does not exist if nobody is looking at it." Einstein was a strong defender of realism in science. He could not have said it more clearly, "The belief in an external world independent of the perceiving subject is the basis of all natural science."

The fact that scientific theories can lead to novel predictions is a strong point in favor of realism. Without realism, there would be no reason to change, correct, or expand scientific knowledge. According to the late Harvard philosopher Hillary Putnam, "realism is the only philosophy that does not make the success of science a miracle." Our minds must have some connection with the things around us, otherwise we could formulate whatever world that we would like to have. If that were the case, we would live in our own separate worlds. Our minds would no longer have any reality check. There would be no grounds left to correct us in what we claim to know.

Another strong point in favor of realism in science is that scientific theories can and often do have consequences that the original proponents of the theory were unaware of. For instance, Fresnel's wave theory of light predicted that light should travel faster in air than in water, whereas the Newtonian particle theory predicted that the speed in water would be the greater. Fresnel's theory turned out to be right. This shows that scientific theories have an objective, realistic structure outside of individual minds. Otherwise, there couldn't even be such a thing as falsification.

Examples of realism in science are plentiful. The planets which astronomers talk about are real objects in the Universe; if they cannot be located—e.g. Vulcan—then they must be phantoms. The closed rings of some organic compounds, such as benzene, were once suggested by the chemist August Kekulé, but can now be made visible through electron microscopes, for instance. As it turned out the ring structure was not just imaginary or make-believe but "real."

Another scientist, the physicist Heinrich Hertz, reported that he had been able to produce the invisible fields of Maxwell's electromagnetic theory in a "visible and almost tangible" form. One of Michael Faraday's favorite demonstrations was showing those "enigmatic" magnetic field lines by sprinkling iron filings onto a sheet of paper held over a bar magnet. The filings align themselves with the magnetic field, thus showing the pattern normally invisible for our eyes, but apparently very "real."

Realism is strongly connected with the search for *truth*. As Peter Medawar once said, "Science will dry up only if scientists lose or fail to exercise the power or incentive to imagine what the truth might be." Science is in fact one of the strongholds where truth is still respected in our culture in spite of an emerging relativism (another one is probably the criminal court system, which demands "the truth and nothing but the truth"). The much heralded idea of falsification says that a theory is in trouble when its predictions turn out to be false—that is, *not* true. This raises the question: what is this truth—how can it be defined and how can it be determined?

St. Thomas Aquinas, for one, holds a very safe and sound view of truth, the same view Aristotle had promoted— sometimes called the *correspondence* theory: "To say of what is that it is and of what is not that it is not, is true." Aquinas himself speaks of a correspondence between reality and intellect (*adequatio intellectus et rei*). This conception may sound trivial, but it basically asserts that any true understanding of reality through the intellect has to correspond to that very reality.

However, the concept of "correspondence" is still ambiguous. It can be taken in the sense of "corresponding *to*" the facts, meaning that a statement is linked to a fact, like a ledger entry in bookkeeping corresponds to a financial transaction. Or it can be taken in the sense of "corresponding *with*" the facts, meaning that a statement conforms or squares with a fact, just like a key corresponds with a key hole. The latter interpretation is harder to defend than the former, for the intellect always processes sense-data in order for perception to become cognition, but let's leave it at that. Compared to two rival theories—the pragmatic theory and the coherence theory of truth—the correspondence theory is arguably still the best definition of truth. Let us see why its rivals fail.

The *pragmatic* theory considers statements true, not because they correspond to what is going on in reality, but because they "work" for us in a satisfactory manner. As long as they work, they are considered "true" in a pragmatic sense. This theory confirms that true explanations are indeed successful, but its problem is that successful explanations are not necessarily true. Ptolemy's geocentrism was highly successful in the eyes of sailors and astronomers at the time, yet it wasn't true. What is even more problematic about the pragmatic theory is that having success depends ultimately on truth in Aquinas's sense, since success implies that a prediction corresponds to what is actually taking place in reality. So that takes us back to the correspondence theory.

The other theory, the *coherence* theory, declares a statement true if it is part of a coherent axiomatic system. However, this would entail that a statement being true in one system may be untrue in another system. Consequently, this kind of truth is self-made, depending on the chosen axioms—which is like saying, what is "true" in chess is not "true" in checkers. Admittedly, this idea may be helpful in mathematics and logic, but it is pretty useless when dealing with reality as done in the empirical sciences. Einstein said it right, "As far as the laws of mathematics refer to reality, they are not certain; and as far as they are certain, they do not refer to reality." When we add 1 to 1 in math, the result is necessarily 2 (so, $1+1=2$); but when we add 1 drop of water to another drop of water, the result is not 2 drops but 1 drop (so, $1+1=1$); and when we add one organism to another, we may end up with three or more (so, $1+1 \geq 3$).

Also, it is completely up to us as to whether we choose either a Euclidean or a Riemannian geometry, but when it comes to cosmology, we need the kind of geometry with the best correspondence between our theories and reality. The problem is that the choice between coherent systems cannot be made on the basis of coherence but must be made on the basis of something else—correspondence, that is. Maurice Merleau-Ponty said it right, "the real is coherent and probable because it is real, and not real because it is coherent." So we are back again with the correspondence theory.

So we must come to the conclusion that truth is ultimately based on correspondence—which is probably something that practically all modern scientists would agree with, although perhaps not wholeheartedly and explicitly. They do in fact maintain, as a working assumption, that they are dealing with objective reality, even though they do not always admit it. At the moment they would assert instead that their beliefs are mere artifacts, such a claim would act like a boomerang that destroys its own truth claims. Beliefs like these contradict the fact that scientific proofs of something being true come from conformity with reality, not from systems of ideas.

As a matter of fact, we claim our knowledge to be real, truthful, objective, and reliable—after being tested versus reality, of course. This viewpoint should not be confused with "naïve realism," which assumes that perceptions based on the senses are direct recordings of reality. If that were the case, all we would need to do to perceive is to take a "picture" of it. However, as we found out, perception is not a passive monitoring process of sense data but rather the result of some intellectual reconstruction process with the help of concepts.

Yet, there seems to be trouble at the horizon for the theory of realism and the correspondence theory. Some claim that embracing a realism that affirms the existence of the objective world and our ability to know it has become outdated and has forever been defeated by the quantum physicist Niels Bohr and his school—in spite of objections made by some giant opponents such as Albert Einstein, Max Planck, David Bohm, and Erwin Schrödinger, to name just a few. What could the problem be?

According to Werner Heisenberg's "principle of *uncertainty*," it is impossible to determine simultaneously the values of a particle's position and momentum (the product of mass and velocity) with any great degree of certainty; the more precisely one property is known, the less precisely the other can be known. The reason is that in order to make an observation, scientists must intervene in the particle they are observing. As a result, measuring a particle's position affects the measurement of its momentum—and reversed.

By the way, this is not unique to quantum effects, for it is a phenomenon also known in classical physics. To take a simple case, measuring the temperature of a fluid with a simple thermometer means that the temperature is affected by the temperature of the thermometer itself. The difference is, though, that this inaccuracy can be corrected for, which does not seem possible in quantum physics.

Does this mean there is no longer an "objective reality" outside us, characterized by particles which occupy a well-defined location and a well-defined momentum? Perhaps it does, but not necessarily so. In other words, we might dealing with a deep-seated "belief" that can no longer be confirmed by quantum physics. In Niels Bohr's interpretation of this phenomenon, which is called the "principle of *indeterminacy*," the values of a particle's position and momentum are indeed and in essence undetermined until we measure one of them. Although most quantum physicists nowadays accept this account, the question is how to interpret this phenomenon of "uncertainty" and more in particular of "indeterminacy." It is striking that there are at least seventeen different interpretations of quantum mechanics. "No one understands quantum physics," the Nobel Laureate in physics Richard Feynman famously said.

Which side is right? Obviously, scientific debates should never be decided by a majority vote. What we need instead is a scientific assessment—although it is not quite clear whether we are dealing here with a scientific or rather a philosophical disagreement. Some claim that the Copenhagen interpretation is actually an attempt to do metaphysics under the guise of physics.

It's too complicated an issue for this book, but a few general comments or critical remarks should be made. Ironically, Bohr and his Copenhagen school try to give us a causal explanation of the alleged "fact" that causal explanations are no longer possible in their view. They think that we do not know because we cannot know. The problem is, though, that our perceptual experiences can only give us knowledge of the external physical world if they are *causally* related to that world. To deny causality in the name of science would therefore be undermining the very empirical foundations of science itself (§14).

How could we ever account for our knowledge of the world that physics tells us about if we have no causal contact with it at all? Furthermore, it could be argued that it is logically impossible to prove that something has no cause at all. No search can ever reveal the *absence* of what it was searching for. Causality can never be conclusively defeated by experiments, since causality is the very foundation of experiments and of the way science works. The physical order we observe in this world appears to be amazingly "consistent," so why would it be different when it comes to quantum events? Those who claim there is a difference owe us at least an explanation as to why we should draw the line where they say we should, and how there could be such a line.

How are we to assess this discussion? Perhaps the comparison with radioactive decay may help. As we will discuss later (§19), the moment in time at which a particular nucleus will decay is unknown and unpredictable, yet a collection of atoms of a radioactive isotope will decay exponentially at a rate described by a parameter known as the half-life. After one half-life has elapsed, one half of the atoms of the isotope in question will have decayed into a "daughter" isotope or decay product. The half-life parameter is something found in "objective reality." This only means that not all laws of nature are deterministic—some are probabilistic. Yet both are part of "objective reality."

Besides, it is not the place of physics to elevate the indeterminacy of a single nucleus' decay to the indeterminacy of nature itself, let alone the existence of an external world. Would particles really not exist until we observe them? At one point, for instance, even geneticists thought that genes were merely mathematical constructs, until it was shown they were real entities in DNA (§4 and §10). The question about the existence of an external world can probably be best answered with Stanley Jaki's statement that the uncertainty principle is not a statement about what the quantum world is like, but rather a statement about what we *know* about the quantum world through observation. In more technical terms, it's an epistemological, not an ontological, issue.

Nevertheless, observations remain vital elements of all empirical sciences. The truth of the matter is that the testimony of the empirical sciences is and remains the testimony of our five senses. The edifice on which the empirical sciences are built rests on the human eye, ear, nose, tongue, and sense of touch. However, science has enlarged the range of our senses with microscopes and telescopes, X-rays and radar, all of which have made the influence of theories on perception and observation even more pervasive. The fact is that most factual material is the product of special observational techniques. Because these observational techniques have a theoretical basis, perception and observation rely heavily on observational theories. Reliance cannot be placed on experimental results unless the experimenters are thoroughly competent and familiar with the technical procedures they use.

What we said so far about science also holds for religion. Both are about reality, about objective truth, and about unchanging facts. We want to know the *truth* about the world and about God's dealings with the world—not in the form of fantasies and imaginations but in the form of *facts*. This seems rather obvious in the case of science, but is sometimes questioned when it comes to religion. Religion is not merely about feelings and sentiments but also about certain claims regarding religious facts. The Nicene Creed, for instance, has a long series of "is" statements; it proclaims religious facts and religious truths.

Since even religion is about facts, it is based on empirical evidence, not to be confused with the experimental evidence that most natural sciences use in addition. In religion we reason from the "seen" to the "unseen," from visible data to invisible facts. This reasoning is done by the use of logic and philosophy. If there is anything inconsistent or contradictory in religion then it cannot be a religious fact.

It must be clear by now that setting facts against interpretations is false. Both science and religion deal with facts— that is, with factual interpretations of things, situations, and events. Scientists must submit their minds to the facts of experiment, whereas religious believers must submit theirs to the facts about God—to how God revealed himself through Scripture, tradition, and reason. In other words, even God's

Revelation as seen in religion reveals real facts about the way God is and deals with us. No matter whether it is in science or in religion, we have to obediently listen to the facts. We cannot force or change the facts to our own liking, for then they are not facts anymore but merely products of wishful thinking.

In general, we tend to overestimate the power of facts when it comes to science, whereas in religion, we tend to underestimate their power. However, in science, there is much more believing in what we know than many want to believe (§15). And, vice-versa, in religion, there is much more knowing in what we believe than many seem to know. Science has theories to help us understand the world better, but they are subject to change—so let us not make science more than what it is. Religion, on the other hand, has truths about God we try to understand more and more, but they never change—so let us not make religion less than what it is.

IV

Where Do Facts Come From?

The title of this section may seem silly at first sight. Of course, one might say, they come from reality, from the world around us—that's where we find them. But that cannot really be true, for facts are not things we can stumble upon or can bump into, so we found out. They are not the kind of entities that we can see, feel, or hear with our senses. We may be able to test them that way, but facts are non-material entities in themselves. We can "bump" into things, events, and situations, but not into facts.

At one point (§8), we provisionally defined facts this way: facts are interpretations by the human intellect of events by means of thoughts and statements. These three elements play an important role: through events, facts can be tested; through thoughts, they can be understood; and through statements, they can be communicated.

From now on, however, we are going to tighten our terminology. Instead of thoughts, we will focus on *concepts*, which were explained earlier (§9), and instead of statements, we will use the more technical term *propositions*, because statements are often seen as merely linguistic entities, which may vary between languages, whereas propositions are more of a logical, philosophical nature. A proposition is the meaning behind a statement. Statements merely express propositions. Hopefully, it will become clear soon why this terminology was chosen, actually mandated.

11. A World of Concepts

There is a lot of confusion about the status of concepts. Some think a concept is a thought, only existing in someone's mind. Others think a concept is a word, only existing in some language. Then there are those who think a concept is a material entity that we can point at. Who is right? Let's find out.

The most common, but erroneous, idea is that concepts are thoughts. But that cannot be true. The reason is basically very simple: If concepts were just thoughts, then we could not communicate with each other, because I cannot read your thoughts, and you can't read mine; I wouldn't know what you are thinking when I use a certain concept for something I am thinking. I would never be able to know you are thinking the same thing as I am thinking when we speak about something like a gene, a mutation, or whatever the two of us wish to talk about. When you and I think about randomness, for instance, it is not that both of us are entertaining our own private thought of randomness, with nothing in common between us. If that were the case, each one of us could have a different thought or idea of randomness, and communication between us would be impossible.

However, when the two of us have the same thought, there must be something we have in common that transcends our private thoughts. That's where concepts come in. In other words, we may think about a concept, but the concept itself is not a thought—only the object of a thought. Concepts do not come from thoughts but they make certain thoughts possible. According to St. Augustine, there must be something "that all reasoning beings, each one using his own reason or mind, see in common."

Another common, but erroneous, idea is that concepts are merely linguistic entities. But that cannot be true either. It is true that words are linguistic items, but concepts are not—they are extra-linguistic entities. So we shouldn't confuse words with concepts. A word is at best a label for a concept. The word "snow" would be a label for the concept of snow. A concept like snow can be conveyed by different words in different languages—for example the word "snow" in English and the word "Schnee" in German.

However, the concept itself would exist even if those languages did not exist or no one had ever used any words to refer to the concept of snow. We can talk about snow found at a time when there were no human beings yet. Claiming that the term "white" in "Snow is white" is just a term, but that there is no such thing as "whiteness" should be countered by asking the question why we apply the term "white" to just the things we do. It's hard, if not impossible, to come up with a better answer than "because they all have whiteness in common"—which brings us back to affirming the existence of universal concepts after all. We may talk about concepts with words, but the concept itself is not a word. Put differently, concepts do not come from language; instead, they make language possible.

Another erroneous, idea is that concepts come from definitions that we make up ourselves to explain what they stand for. But the problem of this is that definitions inevitably require other concepts. Take the following references in a dictionary: "A gene can take alternative forms called alleles," and "An allele is one of the alternative forms of a gene." These are obviously circular references that explain a concept with the very same term it is supposed to explain. But even if the definition is not that trivial but provides a more extensive description, the circularity remains, for every description of a concept requires the use of other concepts, which in turn require further descriptions using concepts.

Dictionaries, for instance, must always ultimately use circular reference since all words in a dictionary are defined in terms of other words. Dictionaries can never step outside their own confines to refer to something outside of the dictionary. No matter how hard we try, we will never be able to get the concept we try to describe and define "off the ground"—every trial falls back on other concepts. Even if we decide to stop this endless regression by declaring a few concepts as the pillars that carry the rest of the conceptual framework, then we still need to explain where those fundamental concepts come from.

Another erroneous, idea is that concepts come directly from observation. Take the concept of hydrogen. It does not arise from particular observations, because no observation can tell us that a certain gas is hydrogen, until we know what hydrogen is and how to test for it. When we see blood stains at crime scenes, the mere observation of those stains doesn't make it blood. Forensic scientists will tell you we need luminal, for instance, to detect blood, as it reacts with the iron in hemoglobin.

But what is even more crucial is this: observation is always about particular things, whereas concepts are universal. Concepts abstract from particular observations that which is universal. Concepts have the universality that observations miss. Material things are always particular, whereas concepts are always universal. We may have seen many circular objects, but we have never seen the perfect circle that the concept of circle is about. Even the concept of uniqueness is universal. It does not apply to one single, unique case, but puts a thing into the category of things that are unique in one or more respects. It's a universal concept about many concrete cases that are similar in a certain aspect—that is, in being unique.

Another erroneous idea is that concepts are established by pointing at something. First of all, for certain concepts there may be nothing to point at. In order to explain the concept of "tomorrow," for instance, there is nothing to point at (other than on a calendar, but that requires the concept of calendar as well). Neither does the mathematical concept of π (pi) refer to any object in the world. There are no π objects.

Second, pointing at a cell under the microscope does not generate the concept of cell. Even showing or pointing at many cells does not generate that concept. Claiming that objects are similar to each other in a certain respect invokes another abstract and universal concept, "similarity," which leads us into a vicious regress problem. In order to decide which objects are to be included in the set and which not, we need a criterion which says that only cells are to be included in the set. This account presupposes the very concept of cell, the acquisition of which it is meant to explain. It's only when we know the concept of cell that we can point at a cell and identify it as a cell. Anything can be pointed at, once it has been identified; but not everything that has been pointed at, can be identified with a concept. As St. Augustine once put it, "what is recognized is present in common to all who recognize it." There is no recognition without cognition.

Another erroneous idea is that concepts are material entities themselves, located *outside* the mind. But this idea cannot be true either. As we said earlier, material things are always particular, whereas concepts are always universal. The concept of circularity, so we found out earlier, is not about one particular object with the shape of a circle, but instead it refers to *all* objects with such a shape. The concept of circle can be used for any specific circular object regardless of its size and its imperfections. Concepts have a universality that material objects can never possess. Therefore, there is no material entity outside the mind that the concept of circle refers to.

Besides, even if a concept may sometimes refer to a material entity, the concept itself is not a material entity existing outside the mind in the world around us. Concepts don't have material properties—they have no weight, no shape, no color. Therefore, the concept of circularity is not circular itself; the concept of electron is not an electron itself; the concept of gene is not a gene itself. Particular entities in the world around us, outside our minds, are material, whereas concepts are immaterial.

Then there is this rather common idea that concepts are material entities located *inside* the brain. But this idea cannot be true either. To explain why not, we may need some more explanation, because it is a rather popular misconception among scientists.

It seems to be very tempting for scientists to get rid of the highly abstract and ethereal notion of concepts by reducing them to something that is the product of neural activity. The famous example to defend this position is the following: When a frog sees a fly zooming by, the frog's brain displays a certain pattern of neural firing. In a similar way, when we see a tree, there is a distinctive pattern of neural firing in our brain that is correlated with and caused by seeing a tree. This has led some to believe that thinking of a certain concept—say, a "fly" or a "tree"—is also and only a certain pattern of neural firing. However, the problem here is that seeing a fly or a tree is a matter of perception, whereas using a concept is a matter of thinking. Thoughts have meaning and content, which perceptions do not.

If my thoughts about mathematical circles, for example, were just a physical representation in the form of a neuronal firing pattern somewhere in the brain, those thoughts would at best be another particular material thing. But this necessarily means that those physical, material representations could not be universal. Reducing concepts to material entities—for instance, neuronal firing patterns in the brain—would make them something particular. Particular material things cannot qualify as universal. If the thought about circles were indeed a particular neuronal firing pattern, that pattern itself would have to be a circle as well—which is obviously nonsense.

Besides, to reduce concepts to a "product of neurons" obscures the fact that "neuron" is also an abstract concept in itself. That would make for a pernicious vicious circle: the very idea that concepts are nothing but neurons firing is itself nothing but neurons firing. Those who claim that concepts are merely products of neurons should realize that talking about neurons requires the immaterial concept of neuron to begin with. In other words, concepts do not come from neuronal activities; instead, we can only understand neuronal activities and talk about them with the help of certain concepts—the concepts of nerve, neuron, synapse, neurotransmitter, and the like.

It is hard, if not impossible, *not* to conclude from this that concepts are very enigmatic entities. This can be best illustrated with a mathematical concept such as π. As Stephen Barr points out, it is not some private experience, like a toothache; it is not a material object like a melon; it is more than a sensation, a neurological artifact, or a genetic product; it is certainly more than a certain pattern of neurons firing in the brain. But above all, it is not even a property of material things, because there are no entities in the physical world with the shape of π. Barr again, "One can have 4 cows, but one cannot have π cows; and one can have a 4-sided table, but not a π-sided table." Yet, we know the meaning of π, although π does not have any concrete reference in the physical world.

This leads us to an even more general conclusion. If intellectual activity with the help of concepts is indeed immaterial, then the thing which carries out that activity would itself have to be immaterial too. How could an entirely material thing like the brain ever give rise to immaterial entities such as concepts? Concepts are not something that the brain secretes like the liver secretes gall. This seems to imply that the mind must be different from the brain. The brain works with neurons, but the mind works with concepts. They are apparently not identical and cannot be reduced to each other.

12. A World of Propositions

When we talk about facts, we do so in *statements* or sentences such as "gravitation varies with the square of the distance." So, why do we need then another term— *propositions*?

Well, the main reason is that statements are usually seen as linguistic items, whereas propositions are extra-linguistic entities. A proposition may be expressed in a statement or sentence, but the proposition itself is not a statement or sentence, so it is not a linguistic entity. The proposition "Snow is white" can be conveyed by different sentences in different languages—but the proposition remains the same. The previous proposition would even be true if those languages did not exist or if no one had never uttered that statement or sentence. So, the fact that sentences are products of convention does not entail that propositions are the products of convention too.

It is impossible to talk about facts in a proposition without using concepts. Put differently, propositions hinge on concepts—concepts are their "building blocks." The simple fact that snow is white can be expressed in the statement "Snow is white," but its meaning is a proposition which uses the concept of "snow" and the concept of "white." A more complicated fact would be the fact expressed in the statement "All organisms contain one or more cells," which is based on a proposition that uses the concepts of "organism" and "cell." All these propositions are abstract entities, not concrete sentences spoken or written in a certain language at a certain time and place. That's why all people can mean the same thing when they say, in whatever language, that snow is white.

Why is the distinction between statements and propositions so essential? Here is why. If they were the same, we would run into contradictions or paradoxes—which indicates there is something wrong here. The correspondence theory we talked about earlier (§10) claims that the sentence or statement "The snow is white" is true if it corresponds to the fact that the snow is indeed white. However, this could lead to a paradox. Think of this simple paradox: a card, on one side of which is written, "The sentence written on the other side is true," while on the other side is written, "The sentence written on the other side is false." One could easily arrive at the paradoxical conclusion that either sentence on the card is both true and false at the same time.

The logician Alfred Tarski showed us how to avoid a paradox like this. He told us to decide whether the sentence that was written on either side of the card is within the language system being talked about or instead within the language system in which the talking is being done. If both sentences are taken to be in the language system that is being talked about, then they cannot also be taken as referring to each other. So in situations like these, we have actually two expressions here: One is a sentence or statement, the other one is a proposition. Given this distinction, we can now say that the sentence "The

129

snow is white" is true if it corresponds to the proposition that the snow is white.

Another reason why sentences must be distinguished from propositions is the following. Propositions are the truth-bearers—they are either true or false. They are distinct from the different sentences we might use to express them. "Snow is white" and "Schnee ist Weiss" are different sentences in different languages, but they express the same proposition. That's why two people can have the same thoughts, even when they use different languages with different sentences. You can make a sentence visible by writing it on paper, but the proposition it expresses is invisible. Like concepts, propositions are abstract objects—objects of thought, yet independent of thought.

We do know that sentences are "real"—they can be spoken, heard, read, or written. But how "real" are propositions? There are many reasons why propositions and other objects of thought are something very real. The most important argument is that propositions do not depend on human brains, or even minds, for their existence. The facts contained in propositions exist on their own, even when no human being is aware of those facts. The proposition *snow is white* is true, regardless of any human being considering it or expressing it in a statement; it was already true even before humanity appeared on Plant Earth. The proposition *snow is white* is true even if no one had ever thought about it or had expressed that proposition in a sentence. When two scientists are thinking about the law of gravity, they are both thinking about one and the same truth—which is the truth of the proposition *gravitation varies with the square of the distance.*

Here is another argument for the independent reality of propositions. The proposition *the three angles of a triangle add up to the sum of two right angles* was already true before any human being had talked about triangles. It was true before any human mind existed. It would even be true if the material world would go out of existence. There is actually nothing material about triangles, for there is no geometrical object in the world with the perfection of a triangle, as we saw before. Yet, there is something "real" about the Euclidean proposition regarding the sum of two right angles in a triangle.

A fourth argument for the independent reality of propositions is the following. Propositions can express something that did *not* happen in the material world. Take the proposition *Charles Darwin never read a paper written by Gregor Mendel.* That proposition is an abstract object that cannot be linked to any concrete event. Yet, the object of this thought is a reality, albeit a reality independent of any concrete event. In short, it's a proposition.

A fifth argument comes directly from science. Scientific propositions reflect abstract objects that are independent of us and exist even without us. The law of gravity, for instance, was already operational as an abstract object before human beings came along, and it was operational already before Isaac Newton introduced and discovered gravitation. But what is more, we can think about possible worlds the way scientists do with their hypotheses. We can even think about possible worlds in which the laws of nature are radically different from those that are actually operational right now. Possibilities like these exist in a way independent of us, but certainly independent from the actual material world we live in.

This last argument opens up the possibility that we might be dealing with different "worlds" of propositions. As a matter of fact, propositions can be about actualities, possibilities, or necessities. It is the world of *actualities* that we live in and science deals with. It is the world of *possibilities* that we love to dream about, fantasize about, hypothesize about, and write about in science-fiction. It is the world of *necessities* that we are locked into. Facts, obviously, belong to the world of actualities. Hypotheses and theories belong, at least in origin, to the world of possibilities, until science—or religion, for that matter—finds out which possibilities are in fact actualities. Logic and mathematics operate in the world of necessities, so logical and mathematical truths are propositions about necessities.

The world of *actualities* is probably what we are most familiar with. We love the facts and are constantly in search of the facts—which we typically take as actualities, until the fact turns out not to be a fact. Less familiar are we with the world of *necessities*, as we find them in logic and mathematics. It is a necessity, for instance, that everything is identical to itself, that contradictions are impossible, that 2+2=4, that the sum of the angles of an Euclidian triangle is 180°, and that the series of mathematical numbers is infinite. Propositions like these are necessarily true. We can't change them to our liking.

And then there is the world of *possibilities*. It harbors possible concepts such as Atlantis, Vulcan, phlogiston, centaur, and unicorn—concepts which could be actual but are not. Things could have been different than they actually are, at least in principle. There could possibly be unicorns in this world, but there are actually none. The world of possibilities harbors not only possible concepts but also possible statements, usually in a conditional form. The laws of nature, for example, could have been different from what they actually are. The "multiverse" theory in cosmology even assumes there are multiple universes, of which our own Universe is only one.

Here are some other examples: "If the law of gravity had not been an inverse square law, then the Earth and the other planets would not be able to remain in stable orbits around the sun." Or "Had the strong nuclear force been weaker by even as little as 10 percent, it would not have been able to fuse two hydrogens together to make hydrogen-2 (deuterium), and the prospects of life would have been dim indeed." Or "Had the electromagnetic force been much smaller—say, only one fifth of what it is now—then there could only be some twenty-five elements in nature." Or taken from religion: "If God had not created us, we would not exist." Or "If God is not the First Cause, there wouldn't be any secondary causes." Or "Had Jesus not come down from Heaven, we could not have been saved." Or "Had Jesus not been crucified, we could not have been redeemed."

What all these possibilities have in common is that they are real, but not actual. In other words, the world of possibilities is as real as the world of actualities, not to mention the world of necessities. All three of them contain concepts and propositions—actually in an infinite amount. All together they determine what we do find (actualities), what we must find (necessities), and what we might find (possibilities).

13. The Divine Intellect

If it is true—as we concluded so far—that concepts and propositions don't come from our thoughts, nor from our observations, nor from our languages, nor from our surroundings, nor from our brains with their neurons, then we should ask ourselves from where else they could come. If propositions and concepts are real, where do we find them? Where do they reside? They *must* exist somewhere for them to be available to each one of us and all of us together. But where is that "somewhere"?

If abstract objects such as concepts and propositions do not depend for their existence on the material world nor on the human mind, then there is only one rational option left: they must exist in a "third realm" that is neither material nor mental. This idea is usually associated with the philosopher Plato, who mentioned this third realm long ago. However, Plato's position faces multiple, rather technical problems, which we will not discuss here. But there is a much more acceptable version of this third realm—arguably the only valid one—which goes basically back to St. Augustine (especially in book 2 of *On Free Choice of the Will*) and was later elaborated on by the philosopher Gottfried Leibniz (especially in sections 43-46 of the *Monadology*). This version holds that abstract objects do indeed exist, but they can only do so in an *infinite, eternal, divine intellect.*

Why can't abstract objects exist in *human* intellects, instead of one *Divine* Intellect, you might ask. The reason is that human intellects are contingent—they do not have to exist, but come into being and pass away. If abstract objects only existed in human intellects, they would have to come into existence and could go out of existence too. In addition, we could not have them in common with other human intellects. And more importantly, before humanity emerged, there would and could not have been any abstract concepts.

So the only sort of intellect on which abstract, universal, and timeless concepts and propositions could ultimately depend for their existence would be an intellect which could not possibly have *not* existed—which is a Divine Intellect, the Mind of God, the First Cause. It is this Divine Intellect that grasps and holds all of the logical relationships between all propositions with all their universal concepts—a Being that eternally understands all actual truths, plus all possible truths, as well as all necessary truths.

So the only sort of intellect on which concepts and propositions could ultimately depend for their existence would be an intellect which exists in an absolutely necessary way—a

Divine Intellect, that is. For example, a proposition such as *it is true that snow is white* exists in the Divine Intellect. A proposition like this will be true or false because God causes the world to be such that this proposition is either true or false. So propositions are true, because they exist as true thoughts in the Divine Intellect.

Not only does the Divine Intellect hold all truths—*all* actual, possible, and necessary concepts and propositions—but also *all* logical relationships between them, which necessarily makes for an infinite number of them. Needless to say that this entails an all-knowing Being, God. It should not surprise us then that this conclusion has been called an "argument from eternal truths" for God's existence. Again, it's not a proof based on physics but on metaphysics, for God is not a material entity but rather the ground and source of all that exists in this world. God is the necessary cause of non-necessary, contingent beings, including concepts and propositions.

God is the First Cause of all secondary causes that we find in the world, as we found out earlier (§5). The use of reason can show us in no uncertain terms that the only explanation of all the causes science deals with has to be found in an ultimate First Cause, which makes all secondary causes possible and causes them to be causes on their own. Similar to what the "First Cause argument" does, the "eternal truths argument" proves that the truths of concepts and propositions can only exist within a Divine Intellect, and could not exist if God did not exist.

Although the "eternal truths argument" is in essence a philosophical proof of God's existence, it is surprisingly in line with what religion says about God. The Judeo-Christian faith tells us that we were made in God's likeness and image (Gen. 1:27). Since we were made in God's image and likeness, our finite human intellects are a reflection of God's Divine Intellect and have some kind of access to God's infinite Divine Intellect. Our human intellect is somehow able to capture concepts and propositions which are residing in God's Divine Intellect. Only God's Divine Intellect can make it possible that you and I share the same concepts and propositions when we say and think that snow is white. And the same holds, of course, for all the other concepts and propositions that we entertain, including the ones from science and religion.

To put it in a more charged way: without faith in God's Divine Intellect, we have nothing to claim as truth. We are only entitled to say that the statement "snow is white" is true if snow is indeed white in the mind of God. One of the ways to find out whether it is indeed in the Intellect of God is by "reading God's Mind" in nature and through reason. We do so, for instance, when we "interrogate" the Universe by investigation, exploration, and experiment, but also by using logic, reason, and philosophy.

The existence of a Divine Intellect would also explain why there seems to be a rather perfect "match" between the rationality of our minds and the rationality found in the world around us. Somehow the human intellect seems to be able to capture reality the way it *is*. Not only does the physical order we observe in this world appear to be amazingly "consistent," but so does the world of thoughts in our minds. It is a consistency that must perplex us.

How is it possible that reality can even be "grasped" by our intellect? The mystery we have here is the fact that the rationality present in our intellect matches the rationality we find in the world. The fact that the Universe has an elegant, intelligible, and discoverable underlying mathematical and physical structure calls for an explanation—or otherwise is left unexplained. Even scientists uphold the conviction—consciously or subconsciously—that there is an intelligible plan behind this Universe, a plan that is accessible to the human intellect through the natural light of reason and human exploration.

Where does this power of reason in the human intellect come from? As the physicist John Polkinghorne puts it, "Such a reason would be provided by the Rationality of the Creator." Only the Rationality of the Creator can explain that the world is an objective and orderly entity investigable by the human mind because the mind too is an orderly and objective product of the same rational and consistent Creator. Fr. George Lemaître once spoke about the God of the Big Bang as the "One Who gave us the mind to understand him and to recognize a glimpse of his glory in our Universe which he has so wonderfully adjusted to the mental power with which he has endowed us."

Later on, the late astrophysicist Sir James Jeans would put it this way, "[T]he Universe begins to look more like a great thought than a great machine." The "great thought" that Jeans speaks of in this quote is not just a thought of the human intellect but rather a "thought" of the Divine Intellect. Apparently, there is an intelligible plan behind this Universe, a divine plan that is accessible to the human intellect through the natural light of reason. It is in the Divine Intellect that facts exist, scientific as well as religious facts. Facts are the content of propositions, and like all other propositions they reside in the Divine intellect. That's also where their truth comes from.

As Karl Popper once wrote, "Knowledge in the objective sense is *knowledge without a knower*; it is *knowledge without a knowing subject*." Popper was right when it comes to the knowledge of individual people, whether they are scientists or not. But he was wrong in another sense: even without a human knower, it is still objective knowledge and thus must exist somewhere outside the human mind—that is, in the Mind of God. God is the "knower," the "knowing subject" behind all our scientific and religious knowledge. There cannot be any knowledge without the knowing subject of God. This explains why scientists can do their work of exploring and investigating nature. What they are actually doing in their research is reading the Mind of God—often without them even knowing it. The same holds for religion.

Why do philosophical proofs of God's existence—persuasive as they are—not convince everyone? Why doesn't everyone acknowledge their conclusive power? Well, what we often see in the lives of non-believers, agnostics or atheists, is a certain disconnect between what they know and what they prefer to know. St. Thomas Aquinas was very aware of this disconnect: "Whereas unbelief is in the intellect, the cause of unbelief is in the will." No matter how strong the rational evidence is in favor of God's existence, some choose not to accept God's existence as a fact because they don't like the way the world looks to them with God in the picture. They act like people who deny the holocaust and ignore the overwhelming evidence because they are not willing to believe it. It is the will, says Blaise Pascal, which "dissuades the mind from considering those aspects it doesn't like to see."

V

No Science without Religion

In this section we are going to explain why science cannot really exist or even achieve what it is trying to achieve without religion and without the existence of God. In other words, science cannot really "live" without religion, no matter how strange it sounds at first sight.

One of the reasons is the fact that science needs some assumptions before it can do its work. Although those assumptions often remain hidden or unspoken, without them science would collapse, or could not even get off the ground. Another reason is that the Judeo-Christian religion has played a pivotal role in the emergence of science in the West. The case could be made that science could only get off the ground because of its Judeo-Christian basis.

These issues take us back again to the rather perfect match there is between the rationality of the human intellect and the rationality of the world's structure. It is a match rooted in the Divine Intellect, so we found out. This is especially true of the order and intelligibility we find in this world. The world turns out to be very orderly and very intelligible. Without this conviction, science would be impossible and religion would be unreliable. But there is more to it.

14. The Assumptions of Science

The truth of the matter is that science cannot be done without a series of *assumptions*. Many scientists may not be aware of those assumptions, for they are "hidden" like most of an iceberg is hidden under water. The biologist Ernst Mayr speaks of "silent assumptions that are taken so completely for granted that they are never mentioned." Yet, they play a fundamental role in science. They have to be there before science can even get started, for without them science could not get off the ground. They are proto-scientific, in the sense that they must come first in order for science to follow. So the question is: which are those assumptions? And then there is this more intriguing question: does religion have anything to do with them?

Assumption #1: There is a real world outside of us

This assumption is about a world outside of us and independent of us, open to scientific discovery—which we called *realism* before (§10). This may seem an obvious assumption for scientists, but it has been questioned many times—questioned not only by some quantum physicists but also by skeptical philosophers such as David Hume and Immanuel Kant, who think we can't really know anything about the "real" world behind our observations. Some philosophers have even said that the entire world exists only as a dream in people's minds.

Should scientists take these skeptics seriously? Amusingly enough, scientists sometimes joke about their work with warnings like, "Don't touch anything in a physics lab," or "Don't taste anything in a chemistry lab," or "Don't smell anything in a biology lab." But perhaps they should add also this warning, "Don't trust anything in a philosophy department." Scientists have indeed much reason not to trust those skeptical philosophical views. If there is no real world, then science becomes a hallucination too. So to keep science from crumbling, it needs the assumption of a real world. Without the assumption of a real world outside of us and independent of us, there is no way anymore to distinguish fact from fiction, realities from illusions, and opinions from truths. If there is no objective truth, we are free to believe whatever we like, including utter nonsense. Once scientists give up on the assumption of a real outside world, and of the truths that come with it, they undermine their own findings, changing science into science-fiction.

No wonder Albert Einstein always protested vehemently against such an outcome, as we saw earlier: "The belief in an external world independent of the perceiving subject is the basis of all natural science." Nevertheless, there have been some nonconsenting voices in the field of quantum physics, as we discussed earlier (§10). The quantum physicist Niels Bohr, for instance, seems to be telling us that reality does not exist when we are not observing it—as if the moon does not exist if no one is observing it. But at least he acknowledges that there is indeed a real world that we can observe, but only when we decide to. Besides, one could ask how we can measure something that doesn't exist, so it must have come into existence at least before our measurement, and not because of our measurement.

Why does or should science still hold on to its truth claims? Well, if truth were at the mercy of some individuals, science had to abandon all its universal claims. It is the real world that often forces scientists to revise their theories in order to come closer to the truth. To use an analogy: who would ever want to drive across a bridge designed by engineers who believed their calculations are merely based on opinions instead of truths? Besides, if their claim entails that our beliefs are mere artifacts or illusions, such a claim would act like a boomerang that destroys its own truth claims as well. To know is to know things in the real world, not to know mental abstractions in the mind. Those who say that the world of science is merely an illusion produced by our brains should seriously be questioned about their claim, if this claim would indeed just be an illusion by its own verdict.

So this raises the question of how we know that what we know about the world is about the real world? The short answer is: we don't. We can never prove that we're not all hallucinating, or simply living in some kind of computer simulation. So we are dealing here with an assumption that science cannot prove on its own. Some people think the existence of mind-independent objects can be proven by giving a mighty kick to a stone. But objects perceived by our senses are not the same things as truly existing bodies. Even in a dream, we may have similar experiences, and yet they are mere illusions.

Therefore it does not follow that an objective external world does exist—for we could in fact all be living in a permanent dream world. How would near-sighted people know that the world is not as blurred as they see it? Certainly not by comparing their own images with the "real" images. Perhaps corrective glasses may help them to see better. But at the moment we acknowledge this, we end up talking again about the existence of an objective, real world, distinct from our minds. That's where the buck stops.

Assumption #2: There is order in the Universe.

Not only is science a very orderly enterprise in itself, but it also reveals a very orderly Universe. Interestingly enough, the Greek word "cosmos"—a synonym for the Latin word "universum"—means "order." The role of hypotheses and theories and laws of nature is to capture the very order that would be eluding us if we didn't have science. Thanks to science, we see a striking order in the world around us—for example, planets moving around the Sun in very orderly orbits, with all these orbits lying in almost the same plane, thus making the solar system look like a giant platter.

It is this kind of order that science is in search of and keeps exploring. Take, for instance, one of the oldest examples of a law of nature, phrased by Archimedes (c. 250 BC) as follows: "Any object, wholly or partially immersed in a fluid, is buoyed up by a force equal to the weight of the fluid displaced by the object." It expresses some kind of underlying order in the Universe.

Science reveals something astonishing about this Universe: namely that it's not a chaotic entity. Einstein was right when he wrote, "But surely, a priori, one should expect the world to be chaotic." But instead of the expectation of chaos, we expect the world to have a remarkable order, and it appears to do so. The laws of nature which science brings to light are the many patterns of regularity we detect in this order. No wonder, Einstein also spoke of the "harmony of the Universe."

It is tempting to think that order comes out of chaos. Those who think so may refer to the emergence of the Universe from the "chaos" of the Big Bang. Or they point to what happens when the temperature of water is lowered to the freezing point: the chaotically swirling water molecules begin to line up in a striking pattern. It looks like the orderly pattern of ice is coming out of the chaotic movement of molecules. But the truth is that chaos can never create the order found in this world—as little as blindness can create sight. Order cannot come out of chaos.

The popular saying, "Garbage in, garbage out," applies even here.

We cannot even comprehend the word "chaos" without, at the same time, contrasting it to the word "order"—chaos is that which lacks order. We speak of "chaos," when we haven't been able yet to understand or find out the order behind it. When something seems chaotic—for instance, the transmission of genetic traits to the next generation—science will try to find an orderly, systematic explanation as is done in genetics. It is science's task to unearth the order behind our observations. There is no way scientists can say they discovered, after a long search, that there is actually no order behind what they were investigating. If they do say so, they should be told to keep searching for the order that's apparently still eluding them.

Yet it remains tempting to claim that the world starts with chaos. It is true that evolution seems to have a time arrow moving from "what is less complex" to "what is more complex." However, the "simple" is in no way less orderly than the "complex." Something as simple as a snowflake, for example, actually shows a very intricate order. In other words, there is no reason to think that the process of building more order begins with and is being steered by "disorder" or "chaos." The order we see in nature does not and cannot come from chaos, but must come from a more fundamental, preexisting order at a deeper level. Order can only come from order. Order has to be built in for order to come out.

Stephen Barr uses the following example to show this. Shaking jars of variously shaped candies won't create much more order, for there is no order in their shapes. But when shaking a jar with round candies, we do create more order, as round candies have an underlying order—that is, at a "deeper" level"—which allows for a "hexagonal closest packing" structure. Physics requires the round candies to lower their gravitational potential energy as much as possible by creating the geometry of hexagonal packing. It is this underlying order that enables and explains the order of round candies forming the closest packing structure. In other words, the preexisting order inherent in the round candies is greater than the order that emerges after the candies arrange themselves. It is not an order imposed by scientists but an order waiting to be unveiled by them. And it is not an order coming out of chaos.

The order of the Universe is a fundamental assumption for science. When scientists study matter and energy, they see regular patterns of behavior. It is only due to the orderly design of the Universe that scientists can explain and predict—which would be impossible in a chaotic world of disorder and irregularity. In other words, order must come first before science can even get started. Order is not something scientists discovered after more and more successful searches revealed more and more cases of order. There is no way we can prove order by adding more and more confirming cases, like inductivists might think. Order cannot even be falsified by cases where no order was found (yet). In other words, order is not a scientific discovery but rather a philosophical assumption that enables science and propels scientific research.

If science keeps finding that there is order in the world, why do we still call it an assumption? The reason is that when scientists say they have found something chaotic, the search has just begun to find the order that is still eluding them. Mere chaos is unacceptable in science, because science by its nature is in search of order. Whatever may seem chaotic to laypeople will turn out to be very orderly when science progresses. Those who reject any order in the Universe have basically given up on science. If there were no order in the Universe, it would make no sense to search for laws of nature in physics, chemistry, biology, and other disciplines.

What seems to belie all of this is the fact that many scientists recently developed a strong interest in chaos and chaotic systems, as if these could falsify the existence of "law-and-order" in the Universe. They point out that some natural systems can only be described by non-linear mathematical equations with such complex solutions that we cannot exactly predict what the system will do in the near future. Or to take another example, our measurements of all the initial conditions of a particular system (for example, in meteorology) may be too numerous and/or too inaccurate to predict what exactly the outcome would be. However, this is not really chaos, but only appears to be chaos. As a matter of fact, these scientists are still looking for the very order behind seemingly chaotic phenomena. When the weather forecast is off the mark, we do not conclude that the weather is unpredictable. We just do not know enough to be perfectly accurate in our predictions. But that's not chaos!

Assumption #3: There is causality in the Universe.

Then there is this very simple, basic rule used in all sciences: like causes have like effects. In translation: from causes which appear similar, scientists *expect* similar effects; if there are effects, then there must be certain causes involved; if the effects are different, then they must have had different causes.

The general idea behind this rule is that we live in a world of causality—nothing comes from nothing. This assumption is obviously order-related, but there is more to it. You don't have to be a scientist to understand that like causes have like effects. All of us seem to know this rule almost "intuitively." But scientists in particular make it their "profession" to apply this rule methodically. In their research, they attempt to unveil the causality patterns in the world around us.

But there is a potential problem here: We do not *see* causality—similar to the way we do not see gravity. It's true, we do not see causation in the same way in which we see colors and shapes and motion. So what is causality then? Is it merely something we imagine in our minds? The skeptical philosopher David Hume went that road. Since the supposed influence of a cause upon its effect is not directly evident to sense observation, Hume concluded that the connection between cause and effect is not an aspect of the real world, but only a habit of our thinking as we become accustomed to seeing one thing constantly conjoined to another. Thus Hume reduced causality to correlation at best—no longer as something real found in the world outside ourselves, but rather as a way of thinking about the world.

Does Hume have a point here? His view has certainly not been embraced by great physicists such as Max Planck and Albert Einstein, who all assume that physical laws describe a reality independent of ourselves, and that the theories of science show not only how nature behaves but why it behaves exactly as it does and not otherwise. Besides, Hume's analysis would erase the important distinction between causation and correlation, by having them both reduced to a series of mere subjective associations or generalizations.

In contrast, scientists always want to make sure that causality is not just a matter of correlation. There certainly is, for instance, a correlation between wind velocity and windmill activity, but the correlation doesn't go in both directions—it is only the wind that causes windmill activity, not the other way around. Because policemen are invariably seen around the scenes of criminal acts, that doesn't mean they are causing them. Close relationships are not always cause-and-effect relationships. When Louis Pasteur found a strong correlation between the rate of fermentation and the number of micro-organisms, he had to prove that the microorganisms are not the effect but the cause of fermentation by showing that fermentation does not occur if the entry of microorganism is prevented.

So, where did Hume go wrong? When we see the sun rise every morning, for instance, we know that there is not a different sun rising every morning. Although we know the world through sensations or sense impressions à la Hume, the truth of the matter is that they are just the media that give us access to reality. The Scottish philosopher John Haldane put it well when he said, "One only knows about cats and dogs through sensations, but they are not themselves sensations, any more than the players in a televised football game are color patterns on a flat screen." Hume actually came to his view because he took causality as rooted not in the identity of acting *things*, but in a relationship between *events*, assuming that "all events seem entirely loose and separate" and that "we can never observe any tie between them."

Instead it should be argued that the actions an entity can take are determined by what that entity *is*. When one billiard ball strikes another, it sends the other rolling because of the nature of the two balls and their surroundings. The philosopher James Hill explains what this entails: when we know that billiard balls are solid and when we see one ball moving toward another, then certain effects are quite impossible. The moving ball cannot, for example, just pass through the second ball and come out the other side continuing at the same speed; nor can the first ball stop at exactly the same place as the second ball; nor can one of the balls suddenly vanish, and so on and so forth. The qualities of the balls determine the kind of effect that the impulse of the first ball will have on the second. In other words, when we see entities acting, we see causality in action.

David Hume, on the other hand, seriously tried to get rid of laws of nature by claiming that they are just creations of the mind. However, if Hume were right, laws of nature would exist *only* in the minds of physicists, chemists, and biologists. This solution fails to explain the fact that laws actually do hold in the real world. As mentioned earlier, how is it possible for a bridge that has been designed according to the right laws to stand firm, whereas another bridge collapses because its engineers erred in their calculations or used the wrong laws? Would competent engineers really have better mental habits than their inept colleagues? These laws could never hold if they were only creations of the human mind. That is the reason why laws of nature have to be discovered in nature, over and above being invented in the mind.

It is simply not easy, if not impossible, to get rid of the laws of nature. They describe real patterns of causality in the real world. Since scientists assume that like causes produce like effects, they are able to explain and predict what is going on in the real world. In other words, causality is not something we imagine in our minds but something we assume to be operating in the real world. Anyone coming up with an exception to the rule of causality needs to search further and better, rather than abandoning the rule. The general statement "Like causes have like effects" admits no exception. It is the assumption that nothing happens without a cause.

One could even argue that it is logically impossible to prove that something has no cause at all, since searches like these never reveal the *absence* of their object. Causality can never be conclusively defeated by experiments since causality is the very foundation of all experiments. Since science can never prove there is causality in this universe, it must *assume* there is. Science cannot get started without that assumption, and that very assumption can never be falsified.

Assumption #4: The world is comprehensible to us.

Albert Einstein used to say that the most *in*comprehensible thing about the universe is that it is comprehensible; he actually spoke of a mystery. It is indeed incomprehensible that we can, at least in principle, comprehend the Universe the way it is. It could as well be a complete enigma to us—but fortunately, it's not.

The idea that this Universe is comprehensible, or intelligible, certainly does not come from science itself. Scientists assume and know that the world can be understood and taken as intelligible—otherwise there would be no reason for them to pursue science. They don't know yet completely what this intelligible world looks like, but they do know the world must be intelligible in principle. So intelligibility is definitely not the outcome of intense and extensive scientific research; it is again a notion that must come first before science can even begin. It does not have to be confirmed over and over again, but it is a precondition for confirmation to work. We just *know* that this world is intelligible.

This knowledge is so basic to science that it easily eludes scientists. If you were told some scientists had discovered that certain physical phenomena are not intelligible, you would, or at least should, tell them to keep searching and come up with a better hypothesis or theory—based on this fundamental philosophical knowledge that says the Universe is "fundamentally" intelligible and comprehensible. Science calls for this assumption. It is only because of their trusting that nature is comprehensible in principle that scientists have reason to trust their own research.

That which makes the Universe comprehensible is the existence of concepts and propositions that we extensively talked about earlier (§11, §12). It is thanks to them that we can study and think about electrons, atoms, molecules, cells, and neurons. Concepts make the world intelligible for us and allow us to use them when talking to other scientists (and others). They make the world comprehensible to the human mind and ultimately accessible to everyone. As said earlier, concepts play a central role in how we know the world by helping us see similarities that were not visible before we had them. And scientific propositions, for their part, depend on concepts like these. Propositions about nuclear particles, genes, and neurons remain incomprehensible for people who do not know the concepts used in those propositions.

In other words, the world is only comprehensible and intelligible for us thanks to concepts. Yet, that fact in itself is a mystery, to use Einstein's wording. We mentioned already that this "mystery" can only be understood if there is indeed a Divine Intellect where concepts and propositions reside (§13). Otherwise, even facts would be non-existent, for they are the content of propositions which must reside in the Divine Intellect, from where they receive their truth.

Assumption #5: The world can be put to the test.

We said at the beginning of this book that scientists explore their field of study by using not only their "(aided) senses" but also their "(armed) hands" to manipulate their objects and their "(computerized) tools" to simulate their models. Indeed, most sciences are of an experimental nature. In a simple setting, scientists start with two variables, of which the independent variable is the one freely chosen and manipulated by the scientist in order to trace its effect on the other, so-called dependent, variable.

For modern minds, that is easily understood and actually expected. Nowadays, we no longer accept that science should be done from behind a desk the way it's done in a formal science like mathematics. And our experimental hands would be itching if they were not permitted to touch and manipulate nature. Nevertheless, the attitude and approach of most scholars, centuries ago, were quite different—that is why we call them scholars rather than scientists. What comes with science nowadays did not always come with "science" in the past.

In Aristotle's time, many Greek philosophers considered science to be theoretical knowledge—as distinct from the practical skill of people such as Archimedes. Their stand was that knowledge about nature could not possibly be acquired by disturbing the delicate harmony of nature through what they considered "un-natural" interventions from without. In their eyes, an empirical approach would be incompatible with an experimental approach. Their reasoning was as follows: how can observation in an un-natural experiment be true to nature? By bending nature to one's own will, one could never discover its true features.

That is why many Greek scholars at the time detested experimental interventions. Experiments were not supposed to be part of "real" science; at best they belonged to the field of artisans, where technical skills flourish as a form of art. Scholars usually remained rather distant from artisans. What could barely pass was the anatomist's lancet as a means to remove what obstructs the scholar's view. Aristotle used similar means during his rather accurate observations of the embryonic development of chickens. But in general, theoretical understanding and experimental interference could supposedly not go together. An exception was Alexandria where Egyptian technicians and Greek theoreticians worked together.

It was Roger Bacon (1214-1294) who, together with Robert Grosseteste and St. Albert the Great, clearly articulated the "modern" conception of (natural) science; in doing so, they were the forerunners of the era in which so-called revolutionary science came to life (1300-1650). Roger Bacon introduced the distinction between "passive observation" as performed by the layman and "active experimentation" as done by the scientist. From then on, theoretical understanding and experimental interference were supposed to go hand in hand. Experiments had become an essential part of the "new" science.

This intimate link can also be seen in device-aided observation, once microscopic and macroscopic lenses had made their entry. However, in order to be able to trust new optical devices, one needed some optical knowledge before-hand. That is why the "old-fashioned" scholars did not acknowledge defeat too easily. They questioned the claim that lenses were reliable aids, as lenses may distort reality so much so that they show what cannot possibly exist. These scholars were not yet willing to believe their aided eyes. The conflict came to a head when Galileo showed them his simple telescope (§4). Interestingly enough, Bertold Brecht in his playwright *Life of Galileo* has one of the players in the debate remark against Galileo, "One could be led to answer that your tube—when showing what cannot exist—would not be a very reliable tube, right?" It is a scholar questioning a scientist.

It is clear by now that revolutionary science has won the battle. Observational evidence proved itself more reliable, most of the time, than theoretical speculation. The growth and popularity of this new movement was partly made possible by another Bacon, Francis Bacon (1561-1626). The "Baconian" ideal of science is based on the principle that science grows by gathering empirical and experimental data in an inductive way (§1).

The Baconian ideal of science became gradually implemented in all kinds of institutions—institutes, academies, and societies. A good example of this new era is the Royal Society of

London, which originated in 1660 with a group of thinkers who met to discuss the "new and experimental philosophy" popularized by Francis Bacon. In fact, these were clubs offering a sanctuary to everybody interested in research based on experiments—not only to professionals such as Robert Boyle, but also to amateurs like the lens maker Anthonie Van Leeuwenhoek.

From 1800 onwards, a shift in emphasis occurred, especially at those German universities organized on Von Humboldt's model (1767-1835). In fact, William Whewell did not coin the word scientist until 1840. Science was on its way to becoming more academic and elitist in nature. A new kind of institution pretended to endow academics with scientific "formation," according to a uniform conception of science. After all, due to the influence of Roger and Francis Bacon, all natural sciences had become experimental in principle, and the fact is that experiments cannot be performed without manipulation. Experiments depend on the notion that there is a knowable mechanism linking cause to effect. Experiments work by exerting control over a cause and noting its effects. They basically put nature to the test.

If laws of nature are true, then they are always true. If a certain law of nature is correct and true, then all entities covered by the law must exemplify that law. They may not have a chance to show it, but by experimentation we may give them a chance to show it. Experimentation is no longer seen as something foreign or unnatural to studying nature, but it has become the hallmark of any scientific study.

Conclusion

Let's round up this discussion. Obviously, the assumption of experimental control is closely linked to the assumptions of order, causality, and comprehensibility found in an external, real world. Science itself did not discover any of these. Instead, all of them combined are needed to make science possible the way we know it now. Put in general terms, a science free from any assumptions simply does not exist. They are in essence the pillars on which the "science building" rests.

Without the assumption that there is an objectively real world, science would not be possible. Without the assumption that there is order in nature, science would not be possible. Without the assumption that like causes have like effects, science would not be possible. Without the assumption that the world is comprehensible, science would not be possible. Without the assumption that science can put nature to the test, science would not be possible. In other words, science is something you need to "believe" in before you can practice it.

How do scientists know that this Universe is comprehensible, that there is some underlying order connecting causes and effects, that there is a "law" stating that every event depends on some law of nature, and so on? They cannot prove all of this in an empirical way but must just assume them to be true. If scientists couldn't *assume* all of this, they would have to give up on all their scientific endeavors. Apparently, whatever science can do ultimately rests on what science can*not* do—scientifically testing its own assumptions.

15. Assumptions from Heaven

How do we know that the assumptions which science entertains are warranted and not mere inventions or conjectures? Certainly not through science, as we found out, for science cannot scientifically prove or test its own assumptions—they come before science can even begin. Since science presupposes these assumptions, it cannot attempt to justify them without arguing in a circle. To break out of this circle, their validity and legitimacy must come from somewhere else. From where then? Their definitive source is probably a surprise to many: these assumptions come ultimately from God and reside in Heaven, in a Divine Intellect. How can that possibly be true? Here is how.

If the world is indeed comprehensible, as science assumes it is, then the world must be intelligible to the human intellect. Well, it is the *rationality* of the human intellect that makes the world intelligible and understandable to us; it gives us the power to comprehend the Universe through reasoning and to discover truths about this world. It is the rationality of the human intellect that gives us also access to the Divine Intellect—to God, the laws of nature, and the structure of this Universe.

St. Thomas Aquinas calls a human being *animal rationale*— a term that defines us as animals that use reason [*ratio*]. Rationality is our hall mark. It is our capacity for reasoning that sets us apart from the rest of the animal world. Rationality is our capacity to make judgements and decisions guided by reason. In fact, it is rationality that gives us access to the world of truths and untruths—a world beyond our control. Rationality is our capacity for abstract thinking and having reasons for our thoughts, thus giving us access to the "unseen" world of thoughts, laws, concepts, facts, and truths. Weighing evidence and coming to a conclusion are rational activities of the intellect par excellence.

As we noticed earlier, somehow the rationality of the human intellect is able to capture reality the way it *is*. Not only does the world of rationality in the human intellect appear to be amazingly "consistent," but so does the physical order we observe in this world. How is it possible that reality can be "grasped" by the rationality of a human intellect? Somehow there must be "rationality" in the natural world as well. How could the world be comprehensible if its structure—its "law and order"—did not have some kind of "rationality" in it? The world itself must have some kind of rationality, which makes the world comprehensible. The rationality of the human intellect makes it possible for us to use concepts (§11), incorporate them into propositions (§12), derive test implications from them (§3), and then decide on the outcome. Through all of this, we gain access to the reality of the physical world. The mystery we have here is the fact that the rationality present in the human intellect matches the rationality we find in the natural world.

That there is rationality in the human intellect isn't too farfetched for most of us. But is there really any rationality in the physical world around us? Not in the literal sense, of course. But on further inspection, there are many "rational" elements in the structure and order of the physical world. They make the world look highly "consistent"—which is in essence a rational concept. Somehow the order we discover in the world is rational and logical, and that's why our scientific findings and theories can appear as rational and logical too—making the world look like "a great thought," in the words of the astrophysicist Sir James Jeans.

This would not be a surprise to religious believers, but even scientists come more and more to the conclusion that the universe has an elegant, intelligible, and discoverable underlying physical and mathematical structure. The beauty and elegance of the laws of nature and the mathematical equations behind them point to a Divine Intellect who created them. Even the physicist and Nobel laureate Steven Weinberg, a convinced atheist, had to admit that "sometimes nature seems more beautiful than strictly necessary."

Where does the rationality of the world come from? The fact

that the Universe has an elegant, intelligible, and discoverable underlying mathematical and physical structure calls for some kind of explanation—or otherwise must be left unexplained. We said it several times, even scientists uphold the conviction— consciously or subconsciously—that there is an intelligible plan behind this Universe, a plan that is accessible to the human intellect through the natural light of reason. Nevertheless, our universe need not be the way it is, and it need not even exist. Yet, it is there, just the way it is. The answer that there is no explanation for this—things just are the way they are, as some say—is not a very satisfying explanation that is in fact irrational (besides, we just did give an explanation). We need to look for an answer somewhere—certainly not in science.

Now that the order of the Universe and the laws of nature as discovered by science receive more and more attention, and now that it has been found they form a single magnificent edifice of great subtlety, harmony, and beauty, Stephen Barr finds reason to declare, "[T]he question of a cosmic designer seems no longer irrelevant, but inescapable." One could certainly take this as "empirical evidence" for the existence of a Creator God. Even Albert Einstein had to acknowledge, "Everyone who is seriously involved in the pursuit of science becomes convinced that a Spirit is manifest in the laws of the Universe—a Spirit vastly superior to that of man."

More and more physicists are beginning to agree with this. The physicist Paul Davies said at one point, "There must be an unchanging rational ground in which the logical, orderly nature of the universe is rooted." And the British physicist John Polkinghorne speaks of "pointers to the divine as the only totally adequate ground of intelligibility." In other words, the Universe itself is loaded with reason and rationality. Only the Rationality of the Creator can explain that the world is an orderly, rational entity accessible to the human mind because the mind too is an orderly, rational product of the Divine Intellect. The intelligible plan behind this Universe is accessible to the human intellect through the natural light of reason coming from the Divine Intellect.

There is reason in the human intellect and there is reason in the world around us. How come they match with each other so beautifully? The only rational answer is: they both derive from the Reason we find in the Divine Intellect. That's exactly the answer religion gives us: We were created in God's image and likeness (Gen. 1:26a). Only this can explain the power of reason both in the physical world and in the human intellect.

When speaking of the "assumptions" behind science earlier (§14), we might have created the impression that they only exist in the minds of scientists—as if there were something these scientists have decided to take on in order to legitimize their research. In the eyes of scientists, each one does indeed look like an assumption that they happen to believe in. But if that's all there is to it, science becomes a pretty shaky enterprise, resting on man-made quicksand. Only religion can give these assumptions a firm basis—actually a basis in Heaven, in the Divine Intellect. For religious believers each assumption is not just a "conjecture," but a "given," residing in Heaven in the Divine Intellect. This changes each assumption of science from a "conjecture" of the mind into a "given" in reality. Without God nothing has a firm foundation.

Do we have here another proof of God's existence? In a sense we do. Without a Creator God, scientists would fundamentally lose their reason for trusting their own scientific reasoning. If God does exist, however, there is at least an explanation and foundation for the existence of rationality in the human intellect as well as in the structure of the Universe. We could further reinforce this argument by asking the following rhetorical questions: Could there be order in this world if there were no orderly Creator? Could there be laws of nature if there were no rational Lawgiver? Could nature be intelligible if it were not created by an intelligent Creator? Could there be design in nature, if there were no intelligent Designer? In short, could there be any rationality in our intellect and in the world around us without a Divine Intellect in Heaven? The answer to all these rhetorical questions could be very brief: no, there could not. Without God, none of these questions would even make sense.

All of this is connected with the First Cause concept of God that we discussed earlier. No doubt, though, this concept is rather abstract, whereas God in the Judeo-Christian religion is a very personal God—"God of Abraham, God of Isaac, God of Jacob, not of philosophers and scholars," in the words of Blaise Pascal. Can we make the First Cause less abstract and give it more color and detail? Certainly! Let's see how.

On the one hand, we need to stress that the First Cause is not something but *someone*. God must be *more* than a human being, not less. If God were not a person, God would be *less* than we are, which would be contradictory to being God. Hence, God must at least be a Person, but then a Person with infinitely more power, infinitely more knowledge, infinitely more intellect, infinitely more love—having all of these character-istics to an infinite degree and with ultimate perfection. So when we say that God is all-powerful and al-mighty, we should not misconstrue such terms. God's power does not exceed other powers in degree but beyond any comparison. God is an Infinite Power completely unlike our finite powers. God is not a worldly power raised to the zillionth power, but he is an "other-worldly" power—all-powerful, al-mighty, omni-potent.

On the other hand, we need to make clear also that God is not a person like you and I. When we call God a "Person," we have to understand this term in an *analogous* way. God is not a person like you and I, and yet God is a person analogous to a person like you and I. When we say that God is able to "see," this is to be understood neither as a claim that God has eyes like humans do, nor that God is incapable of seeing because he does not have such eyes. What we mean instead is that God perceives in a manner *analogous* to the way humans perceive with their eyes. The Psalmist could not have said it more clearly, "Does the one who shaped the ear not hear? The one who formed the eye not see?" (Ps. 94:9). Indeed, how can God, who made the eye, not be able to see? How can God, who made the ear, not be able to hear? There is no way.

The Bible answers these rhetorical questions quite succinctly when it says that we were made after God's image and likeness (Gen. 1:26a). Of course, God does not hear like we hear or see like we see. God alone is perfect—we are only imperfect images of him. That's why he sees and hears infinitely better than we do. The First Cause is analogous to, but in no way identical to, finite secondary causes like you and me. God is not one of those secondary causes.

You might wonder, though, whether this is really all we can say about God. Obviously, the Bible tells us much more about God than human reasoning can achieve all on its own. The God of the First Cause turns out to be also a God of Love who provides for his creatures and who came down from Heaven through the Incarnation of his Son Jesus Christ. So God is much more than what philosophy and mere human reasoning can tell us. But at least, God is not *less* than what philosophy can reveal to us by using the power of reason which comes from God too. If that's who God is, then we probably want to come as close to this person, God, as possible. St. Augustine could not have expressed it better when he said, "To fall in love with God is the greatest of romances, to seek Him the greatest adventure, to find Him the greatest human achievement."

16. The Judeo-Christian Roots of Science

In the previous chapter, the assumptions of science gave us an opening for the existence of God. In this chapter, we will discuss the reversed: how the existence of God gave us an opening for the rise of science in the past.

The idea that Christianity gave rise to science might be a surprise to many, as it was to Alfred North Whitehead's Harvard audience in 1925 when this famous mathematician and philosopher told them that modern science was a product of Christianity. They were shocked, probably out of mere ignorance. But the idea was not new, and certainly not flimsy. One of the first in more recent history to be aware of the Catholic roots of science was the French physicist Pierre Duhem (1861-1916). When he studied the works of Catholic medieval mathematicians and philosophers such as John Buridan, Nicholas of Oresme, and Roger Bacon, their sophistication surprised him.

Duhem consequently came to regard them as the founders of modern science, as they had in his view anticipated many of the discoveries of Galileo and later scientists. Thus he came to regard the medieval scholastic tradition of the Catholic Church as the origin of modern science. Duhem had to come to the conclusion that "the mechanics and physics of which modern times are justifiably proud [came] from doctrines professed in the heart of the medieval schools." And those schools were undeniably Catholic.

Pierre Duhem did actually path breaking work in the history of science when he showed that the doctrines of the Catholic Church have been a permanent ally of, rather than an obstacle to, the success of the scientific enterprise in the West. He opened the eyes of many for the fact that it was indeed the—so often despised—metaphysical framework of medieval Catholicism that made modern science possible.

Many historians of science and other scholars would later follow Duhem's lead. The sociologist Rodney Stark at the University of Washington, for instance, argues that the reason why science arose in Europe, and nowhere else, is because of Catholicism: "It is instructive that China, Islam, India, ancient Greece, and Rome all had a highly developed alchemy. But only in Europe did alchemy develop into chemistry. By the same token, many societies developed elaborate systems of astrology, but only in Europe did astrology lead to astronomy." So Stark had to come to the conclusion, "Science was not the work of western secularists or even deists; it was entirely the work of devout believers in an active, conscious, creator God."

Another testimony comes from the historian and economist Thomas E. Woods. As he puts it, "The Roman Catholic Church gave more financial aid and social support to the study of astronomy for over six centuries, from the recovery of ancient learning during the late Middle Ages into the Enlightenment than any other, and, probably, all other institutions." No wonder then that many scientists have thanked the Catholic Church for her support. The nuclear physicist J. Robert Oppenheimer—not a Christian himself—had to acknowledge, "Christianity was needed to give birth to modern science." Even someone like the philosopher of science Thomas Kuhn had to say about Europe—without identifying its Judeo-Christian core, though—"No other place and time has supported that very special community from with scientific productivity comes."

Nevertheless, there are still many people left who are not aware of these facts. The historian of science Edward Grant is probably right in stating that the gift from the Latin Middle Ages to the modern world "is a gift that may never be acknowledged. Perhaps it will always retain the status it has had for the past four centuries as the best-kept secret of Western civilization."

Why was the Catholic Church such a fertile hotbed for the emergence of science? In the Catholic mindset, the Universe is the creation of a Divine Intellect (§13) capable of being rationally interrogated. It is this very Judeo-Christian concept of a Creator God that makes science possible. Belief in a Creator God entails that nature is not a divine but a created entity; nature is not divine in itself, only its Maker is—which opens the door for scientific exploration. For science to arise from a religious cradle, the most important condition is a deep belief in a Creator God. In contrast, if the world or nature is considered divine, then one would never allow oneself to analyze it, dissect it, or perform experiments upon it, and all incentives for doing science would be suppressed. But a created world, by definition, is not divine. It is other than God, and in that very otherness, scientists find their freedom to act. It opens the world up for exploration and investigation.

Without this Judeo-Christian belief in a Creator God, we would not be allowed to even "touch" the divine. A rational God has created a Universe that we can rely on with our rational minds, made in likeness of God's mind (§15). The Book of Wisdom (11:20) says about God, "You have arranged all things by measure and number and weight." Hence the only way to find out what the Creator has actually done is to go out, look, and measure—which is a necessary condition for scientific exploration and investigation. It actually requires the "humility" of scientists to wait for and subject themselves to the outcome of their experiments—for science is about a reality outside the human mind, beyond human control.

It could be argued that a tendency toward a different conception of God is part of what distinguishes other religions from the Judeo-Christian religion—which might explain why natural science improved in the West and weakened, or even lacked, within the rest of the world. Because the Judeo-Christian God is a reliable God—not confined inside the Aristotelian box, not capricious like the Olympians in ancient Greece, and not entirely beyond human comprehension as in Islam—the world depends on the laws that God has laid down in creation. Thanks to God's creation, everything else has lost its power, has lost its divine allure. Faith in this one God changes the Universe—once inhabited with spirits, deities, and goddesses—into something "rational," open to further exploration. In the Catholic view, only God is the source where the order as well as the intelligibility of the Universe ultimately stem from.

As stressed many times already, the only way to find out what this order looks like is to "interrogate" the Universe by investigation, exploration, and experiment (§14). The door for science has been widely opened ever since. It is through scientific experiments that we can "read" God's mind, so to speak. It is this Catholic understanding that the world is both good and intelligible to us that laid the foundation for science and for Western society to pass onto successive generations the discoveries that were made.

Pagan cultures, on the other hand, had created a view of the world that inhibited scientific advancement, as they did not view the world as a rational creation. They viewed things as being controlled by many, even whimsical, gods and magical powers. They did not view the world as something that was governed by laws of nature accessible to the human mind and waiting for discovery. Even the biologist and self-proclaimed agnostic E. O. Wilson had to say about Chinese scholars, "No rational Author of Nature existed in their universe; consequently the objects they meticulously described did not follow universal principles."

165

Some have argued, in contrast, that the order of the Universe is not something rooted in religion but rather something discovered by science. However, science can never prove there is order in this Universe, but instead must assume it, so we found out (§14). The powerful tool of falsification (§3), for instance, is in fact based on this very assumption as well: the fact that scientific evidence can refute a scientific hypothesis is only possible if there is indeed order in this Universe. Without the assumption of order, there would not be any falsifying evidence, for the world would be seen as tainted with exceptional litter. When we do find falsifying evidence, we do not take this as proof that the Universe is not orderly, but as an indication that there is something wrong with the specific order we had come up with. Counter-evidence does allow us to falsify theories, but not the principle of falsification itself.

In contrast, if the Universe is *not* the creation of a Divine Intellect which can be rationally interrogated by all human beings, including the scientists among them, then nature remains an enigma ruled by whimsical deities, chaotic powers, or our own philosophical decrees and regulations. Fr. Stanley Jaki used the phrase "stillbirths of science" in reference to the ancient cultures of Egypt, China, India, Babylon, Greece, and Arabia. Their cyclical worldviews—a "cosmic treadmill" in Jaki's words—prevented the breakthrough of science as a self-sustaining discipline. Jaki claimed that science—as a universal discipline where one discovery leads to another, and where laws of nature and systems of laws are established—was born of Christianity. In his own words, "Within the biblical world view it was ultimately possible to assume that the heavens and the earth are ruled by the same laws. But it was not possible to do this within the world vision that dominated all other ancient cultures. In all of them the heavens were divine."

Whereas almost every culture or religion has given rise to

inventions and some form of technology—for you don't have to be a Christian to invent the wheel—science and scientific exploration of the world around us were exclusively nurtured in a culture with a Judeo-Christian tradition. It is hard to believe this was mere coincidence. Although Aristotle, for example, did make a few significant discoveries, his classical Greek culture was unable to maintain and nurture further development. The Ancient Greek world conceived of the Universe as a huge organism dominated by a pantheon of deities, and destined to go through endless cycles of birth, death, and rebirth. These ancient ideas had to be refuted before real science could emerge in Christian Europe. Jaki again: "Within the Greek ambiance it was impossible, in fact it would have been a sacrilege, to assume that the motion of the moon and the fall of an apple were governed by the same law. It was, however, possible for Newton, because he was the beneficiary of the age-old Christian faith."

Something similar can be said about other non-Christian civilizations. When the first Jesuits went to China, they were amazed at the Asians' lack of progress in their understanding of the world and the heavens. These cultures did contribute talent and ingenuity, but scientific enterprise could never arise. In the Muslim world, to take another example, Muslim mystics such as al-Ashari and al-Ghazzali held that reference to laws of nature was a blasphemy against Allah's omnipotence. There couldn't be laws of nature, for they would limit "Allah's freedom to act" whenever he wishes to act. Even the non-religious philosopher Bertrand Russell had to admit that Islamic "science"—while admirable in many technical ways—could only be seen as "a preserver of ancient knowledge and transmitter to medieval Europe." One could add to this that if Muslims did make some scientific discoveries, it was because they had been influenced already by the Christian perception of the world, centuries before Mohammed came along in the 7th century AD. And then, they have always been surrounded by Christian achievements in science.

The history of science is a clear indication of the positive influence and power of the Judeo-Christian Faith. It is revealing that the "scientific revolution" in the 17th century coincided with the period when Christian belief was at its strongest. It was in God that these pioneer scientists found reason to investigate nature and trust their own scientific reasoning. The founder of quantum physics, Max Planck, noticed that "the greatest thinkers of all ages were deeply religious souls." People who come to mind are Nicolas Copernicus, Johannes Kepler, Isaac Newton, Blaise Pascal, Fr. Gregor Mendel, Louis Pasteur, Fr. George Lemaître, and so many others. So if there is any conflict between science and religion, it is largely a conflict between some men and women of the science community and some men and women of religious communities, rather than between science itself and religion itself.

Apparently, science in itself is not the problem for religious faith. Reality tells us that there are atheistic scientists as well as religious scientists. Atheistic scientists are not inherently better scientists than religious scientists, nor vice-versa. They both can be dedicated scientists who believe in the power of the scientific method. But they differ in one thing: the latter keep an open mind and believe also in the power of religious faith, whereas the former close their mind for almost everything that cannot be dissected, counted, measured, or quantified. They differ in recognizing that there is a legitimate territory of religion outside the territory of science (§6). They also differ in acknowledging that science has basically religious roots.

In other words, there is no intrinsic conflict between science and religion. A scientist who testified to this is the nuclear physicist and Nobel Laureate Werner Heisenberg who once said: "The first drink from the cup of natural science makes atheistic ... But at the bottom of the cup, God is waiting." Something similar was expressed by Max Planck, who revolutionized physics with his quantum theory. It was his observation that "For the believer, God is the beginning, for the scientist He is the end of all reflections."

Interestingly enough, St. Thomas Aquinas had already said something similar: "All our knowledge has its origin in sensation. But God is most remote from sensation. So he is not known to us first, but last." Elsewhere Planck says, "All matter originates and exists only by virtue of a force which brings the particles of the atom to vibration. I must assume behind this force the existence of a conscious and intelligent mind. This mind is the matrix of all matter."

History shows us that the Catholic Church has no fear of science or scientific discovery, in spite of the fabricated myth of a clash between the two. The Church is very definite in her support for the domain of science. This is well put and summarized in the *Catechism* (143): "[M]ethodical research in all branches of knowledge, provided it is carried out in a truly scientific manner and does not override moral laws, can never conflict with the faith, because the things of the world and the things of the faith derive from the same God. The humble and persevering investigator of the secrets of nature is being led, as it were, by the hand of God in spite of himself, for it is God, the conserver of all things, who made them what they are."

It is because of all of this that the Catholic Church has always been very favorable to the scientific enterprise of what we call science nowadays. As a matter of fact, the Catholic Church and the scientific community have a long-standing relationship with each other, actually an existential relationship: without the Catholic Church, there would most likely not be any science. As shown before, science was born in the cradle of the Catholic Church, which might explain why it was not born anywhere else—not in China (with its sophisticated society), not in India (with its philosophical schools), not in Arabia (with its advanced mathematics), not in Japan (with its dedicated craftsmen and technologies), but on Judeo-Christian soil with Judeo-Christian roots.

VI

Religion and Science Need Each Other

The previous chapters show us that science and religion—at least in the Judeo-Christian tradition—have been able to live together, and may even have been favorable to each other. But we could still go one step further: they cannot really live *without* each other.

This conviction has been expressed in various ways by authorities in both science and religion. In the science camp, Albert Einstein used to say, "Science without religion is lame, religion without science is blind." In the religious camp, Pope John Paul II said something similar, "Science can purify religion from error and superstition. Religion can purify science from idolatry and false absolutes." The right-minded representatives of both camps agree that science and religion need each other.

This explains why the two have in fact an existential relationship—one cannot really live without the other. We saw already, for instance, that *science* needs assumptions that can only find their foundation in God (§15). We also discussed that *religion* could have never discovered on its own most of the scientific achievements we are familiar with today (§16). In other words, science and religion offer each other what the other one cannot provide.

17. The Author of Two Books

Another way of saying that science and religion can and should live together can be found in the classic distinction between the "Book of Scripture" and the "Book of Nature." The conception of "Two Books" goes at least as far back as St. Augustine: "It is the divine page that you must listen to; it is the book of the Universe that you must observe." This was in fact a conviction shared in the past by many other Christian thinkers: from early Apologists and Church Fathers to St. Basil; from St. Gregory of Nyssa to St. Augustine, from St. Albert the Great to St. Thomas Aquinas, from Roger Bacon to William of Ockham.

Galileo revived the image of the Two Books. In his Letter to Maria Cristina of Lorraine (1615), Nature and Scripture were presented by Galileo as two books proceeding from the same divine Word. Therefore, he stated about the Book of Nature that the glory of God can also be known by means of the works that He has written in the "open book of heaven." Galileo did mention, though, that this "book cannot be understood unless one first learns to comprehend the language and read the letters in which it is composed." He was referring to the language of mathematics. Johannes Kepler, a contemporary of Galileo, would also speak about the Book of Nature as a book in which God reveals himself in another way than he is revealed in the Sacred Scriptures.

Recently, Pope John Paul's Encyclical *Fides et Ratio* (Faith and Reason) encouraged this dialogue between the "Two Books." And during his October 31st address in 2008 to the Pontifical Academy of Sciences, his successor, Pope Benedict XVI, reinforced the same message when he said, "Galileo saw nature as a book whose author is God in the same way that Scripture has God as its author. It is a book whose history, whose evolution, whose 'writing' and meaning, we 'read' according to the different approaches of the sciences."

What can we learn from the metaphor of the Two Books? There are several aspects to consider.

The first question is: how do we read these two "books"? The shortest answer is: Science knows how to read the "Book of Nature," whereas the Church knows how to read the "Book of Scripture." Somehow, we have two different "authorities" here. Scientists cannot just read whatever they want in the Book of Nature; they need to follow the "rules" of the empirical cycle as followed in the scientific community (§3). Neither can Christians just read whatever they want in the Book of Scripture; they also need to follow certain "rules" of their Church community. In Catholicism, the latter rules are based on Scripture, tradition, and the twosome "faith and reason." Reason makes sure that what has been revealed to us in Scripture is compatible with, and not contradictory to, what we could learn by ourselves. What has been revealed from above and what is understood here below need to be consistent with each other. If not, then we have a problem, for faith without reason is a dangerous undertaking. So faith is not contrary to reason, because claiming that something in our faith was revealed by God requires at least that it's not *against* reason.

Around 400, St. Augustine wrote, "It not infrequently happens that something about the earth, about the sky, [...] about the nature of animals, of fruits, of stones, and of other such things, may be known with the greatest certainty by reasoning or by experience, even by one who is not a Christian." In other words, you don't have to be a Christian to read the Book of Nature. At the same time, Christians should respect what the Book of Nature tells them. St. Augustine once warned

Christians that it is "dangerous to have an infidel hear a Christian... talking nonsense."

On the other hand, science cannot tell Christians how they should read the Book of Scripture either—for that would be intrusion into religious territory. The image of the Two Books creates somehow a protecting fence between the two territories—for, as the saying goes, good fences make good neighbors. Those fences act like a wall that keeps "insiders" in and "outsiders" out.

The second question is: what is the relationship between the Two Books? We know what separates them, but what is it that unites them? The shortest answer is: They have the same Author, God, who revealed himself in nature as the Creator of this world—which is the account of the Book of Nature—and who revealed himself in history when he guided his chosen people in Israel, spoke through their prophets, and ultimately became man in Jesus of Nazareth—which is the account of the Book of Scripture. Both books are like the two versions of the same story or the two sides of the same coin, and thus they complement each other—like a match made in Heaven, so to speak. Pope Benedict XVI once spoke of "nature as a book whose author is God in the same way that Scripture has God as its author." To set the Two Books up against each other creates a false dichotomy.

Some might object, though, that the image of Two Books is deceiving, for being the Author of Scripture is very different from being the Author of Nature. But that's not quite true. In Christianity, the Scriptures are not seen as something that came down from Heaven, directly from God, in a literal way. It is not a text that should be taken in an entirely literal way as is done in some religions. Whereas Muslims, for instance, insist that the Koran (Qur'an) is a perfect transcript of the angel Gabriel's recitation to Muhammad, the Catholic Church proclaims that the Bible was written by human authors, yet guided by divine inspiration. Pope Benedict XV wrote in his encyclical *Spiritus Paraclitus* that what's written in Scripture "ought not to be described as automatic writing." So seen that way, we may still call God the Author of the Book of Scripture similar to the way God is the Author of the Book of Nature.

The third question is: does each one of the Two Books really deal with truths? We discussed already (§8) that both religion and science—that is, both the Book of Scripture and the Book of Nature—deal with facts (as to how to test whether they are really facts, that's a different story). This means there are scientific, natural truths about nature and there are religious, supernatural truths about God. Scientific truths come from the book of Nature, while religious truths come from the Book of Scripture.

However, these two kinds of truths have a rather different status. Science, on the one hand, tries to reach the truth but has not fully captured it yet. Religion, on the other hand, has received the full truth through God's Revelation (and through philosophy and theology), but has not fully understood it yet; although religious truths will never change, our understanding of them may. Whereas science has *theories* which are subject to change so as to help us better understand nature, religion has *dogmas* which we try to understand better and better, but they never change. Truth is truth, but we may not fully understand the truth yet, neither in the Book of Nature nor in the Book of Scripture. We keep trying to come to a better understanding of both kinds of truth.

The fourth question is: do the Two Books each have their own authority? Yes, but it's "borrowed" authority of course—borrowed from the ultimate Author behind the Two Books. But with their borrowed authority, they received their own limited authority. Because either Book has its own, borrowed authority, religion should be protected from science as much as science should be protected from religion. Science should stay away from religious territory as much as religion should stay clear of science. This pleads for a sound separation of both domains in order to prevent any trespassing. Science should never silence religion, nor should religion ever silence science. They both have their own voice, their own authority, their own "magisterium." Scientists have to patiently listen to what the Book of Nature tells them. And religious believers have to obediently listen to what the Book of Scripture tells them. In either case, we are receivers, not providers.

As a consequence, science can neither endorse nor reject what is beyond its reach—issues such as creation, soul, providence, or original sin—and religion can neither endorse nor reject what is beyond its reach either—issues such as heliocentrism or the Big Bang or evolutionary biology. When using the Book of Scripture to claim that those theories are true or false, religion goes beyond its boundaries, for it uses the Book of Scripture to answer questions about the Book of Nature. Instead, we should always try to find out with scientific tools whether a certain scientific theory is true or not. If it is true, religion will have to accept it, for religion seeks the truth too. If it is not true, present-day science will have to modify itself, since science seeks the truth as well. No wonder then, the Ecumenical Council of Vatican II declared that "if methodical investigation within every branch of learning is carried out in a genuinely scientific manner and in accord with moral norms, it never truly conflicts with faith, for earthly matters and the concerns of faith derive from the same God."

This explains why religion has never declared heliocentrism a dogma, and evolution will never be declared a dogma either, nor will the Big Bang theory. Such issues are beyond the Church's authority. Our salvation does not depend on whether we believe in heliocentrism, evolution, or any other scientific theory. Put differently, salvation doesn't come from science but from religion—the concept of salvation just does not exist in the vocabulary of science. Given this sound stance, we should expect something similar from the scientific community. Ideologies such as scientism, materialism, physicalism, and evolutionism go far beyond scientific expertise and should never be promoted as "dogmas" to be held in the scientific community (§6). Good fences make good neighbors.

The fifth question is: how do we read the Two Books? Reading a book—any kind of book, for that matter—means that the book needs to be interpreted in order to be properly read and understood. Something similar is true of the Two Books. As St. Thomas Aquinas used to say, "[It] is the task of the good interpreter to look, not at the words, but at the meaning." The Book of Nature requires that its concepts are properly interpreted and understood. Even the Book of Scripture cannot be interpreted by itself. According to Acts 8:30-31, Philip ran to the Ethiopian eunuch in his chariot, "and heard him reading Isaiah the prophet and said, 'Do you understand what you are reading?' He replied, 'How can I, unless someone instructs me?'" Even Jesus himself often demonstrated how the Scribes and Pharisees used wrong interpretations, and hence he corrects them by properly interpreting Scripture, thus demonstrating that the Scriptures do not interpret themselves.

Interestingly enough, the "Two Books" each have at least more than one reading. For instance, at the time of Galileo, the Book of Nature had the "reading" of *Copernicus* as well as the "reading" of *Tycho*—only time could tell which one would turn out to be right. In a similar way, the Book of Scripture has the *chronological* reading of the first chapter of Genesis as well as the *structural* reading—only the Church can tell us which one is right. The latter example probably needs a bit more explanation.

According to the *chronological* reading, the six days of creation followed one another in strict chronological order. However, there are several indications that this account cannot be taken literally. First there is the fact that the creation of the sun happens three days after the day/night cycle is established. In addition, the plants on earth were created one day before the sun was created. This is especially troublesome if a creation "day" is interpreted as referring to a longer period of time, since plants need sunlight for photosynthesis. Even St. Augustine was able to acknowledge this: "What kind of days these were, it is extremely difficult, if not impossible for us to conceive." So we can draw the provisional conclusion that Genesis 1 is not meant to be understood as a literal chronological account.

This takes us to the second possibility: Genesis 1 is to be given a non-chronological, *structural* reading. Advocates of this view point out that ancient literature commonly placed historical material in a sequential order according to a particular structure or framework, rather than in strict chronological order. The *Catechism* (337) explains that "Scripture presents the work of the Creator symbolically as a succession of six days of divine 'work,' concluded by the 'rest' of the seventh day." The structural interpretation primarily tells us *that* God created the world—God is the origin of everything—whereas science tells us *how* he does it.

The structural approach is not a new, modern idea. For many centuries, it has been recognized that the six days of creation are divided into two sets of three. In the first set, God separates one thing from another: On day one, he separates the light and the darkness (thus giving rise to day and night); on day two, he separates the waters above from the waters below (thus giving rise to the sky and the sea); and on day three, he separates the waters below from each other (thus giving rise to dry land in-between the waters). Classically, this section is known as describing the work of *division*.

In the second set of three days, God goes back over the realms he produced by division during the first three days, and then populates, or "adorns," them. On day four, he adorns the day and the night with the sun, moon, and stars. On day five, he populates the sky and sea with birds and fish. And on day six, he populates the land (between the divided waters) with animals and humankind. Classically, this is known as describing the work of *adornment*.

That this twofold process does indeed represent the ordering principle of Genesis 1 is also indicated at the beginning and end of the account. At the beginning, we are told that "the earth was a formless void" (Gen. 1:2). The work of *division* cures the "formless" problem, whereas the work of *adornment* fixes the "void" issue. Likewise, at the end of the account we are told "the heavens and the earth were finished [i.e., by division], and all their multitude [i.e., by adornment]" (Gen. 2:1). Biblical scholars have recognized for centuries that these are the ordering principles at work in Genesis 1. So this is not something modern Bible scholars have come up with—we find this idea, for instance, in the writings of St. Thomas Aquinas many centuries ago.

The sixth question is: who determines what the correct reading is? For science, on the one hand, it is the scientific community. Scientists are the explorers of the Universe, so they determine how exploring is properly done. Papers in scientific magazines have to go through a peer review before they can be published. We discussed this issue already in the first chapters.

For religion, on the other hand, the 'review" is done by the religious community, more in specific the Catholic Church. She received, passed on, generated, and spread the Book of Scripture, so she is the only one to determine how it should be read. Scripture is a product of its religious community—that's where its cradle was. St. Paul did not walk around with a copy of the New Testament in his pocket—for there was no New Testament yet. So it cannot be true that the Church is the product of Scripture; it's actually the other way around: Scripture is the product of the Church. That's why the Church determines how to read the Book of Scripture, similar to the way she once told the Ethiopian eunuch how to read that Book.

The seventh question is: can the Two Books be read next to each other by the same person? Yes, they are actually alive together in the minds of many people. As a matter of fact, the person of science and the person of faith are ultimately one and the same person. What we know and believe in science and what we know and believe in religion need to live together in the same human person. In fact, in the modern world, it is no longer possible for a religious believer to ignore the work of scientists, and it is probably fair to say that scientists in turn could and should benefit from the worldview of religious believers. We see repeatedly in the history of science that scientists are guided by philosophical and religious ideas, and that religious believers became better informed by scientists. Apparently, the voice of faith does not threaten but strengthens the voice of science, and vice-versa. Put differently, science and religion are in essence very good neighbors that can't survive without each other. So they should live together in peaceful harmony, integrated and united within the same person.

On the other hand, it's obvious that the world of science harbors both religious and non-religious scientists. In other words, you don't have to be a Christian to be a good scientist—although religious faith might be needed to get started, as we found out earlier (§15). While there are atheistic scientists as well as religious scientists, they both are dedicated scientists who believe in the power of the scientific method. But they differ in one thing: they either accept or reject the validity of a religious outlook. In either case, though, it is not science itself that can decide who is right. Such a decision is a matter of "faith" in itself.

The eight question is: do we really need Two Books rather than one? To put it simply, the answer is Yes, for with only one book, one would get only a part of the truth. We have the right to hear the whole truth instead, and not just the part that science tells us, based exclusively on the Book of Nature, or the part that comes from faith alone, based on the Book of Scripture. Science tells us more than we could know by faith alone, and faith tells us more than we could know by science alone.

Therefore, the classic distinction between the "Two Books" is very enlightening. In order to find out more about God, we need to read the Book of Scripture; to find out more about the Universe, we need to read the Book of Nature. Seeing the world from both the perspective of science and the perspective of religion is something the English theoretical physicist and Anglican priest John Polkinghorne describes as seeing the world with "two eyes instead of one." As he explains, "Seeing the world with two eyes—having binocular vision—enables me to understand more than I could with either eye on its own."

It is obvious that religion on its own does not know anything about molecules, cells, hormones, enzymes, stars, brains, computers, and so on. Religion depends on science for discoveries like these, because it does not have the proper tools to learn more about these scientific issues. As a consequence, religion could easily lose its anchor in this world without the input of science. One could even make the case that religion without science may lead to superstition—astrology, horo-scopes, and mediums. Science can protect religious people from erroneous or dangerous beliefs that merely appear to be religious but are actually superstitious.

It is also obvious that science on its own does not know anything about the ultimate questions that religion is known for. Think of questions like these: Why is there something rather than nothing? Why do human beings exist? Why am I here? Why should I be moral? Why should I bother to search for truth? Why should I protect life? Why is there evil in the world? Why do I do what I don't want to do? Why do bad things happen to good people? Why do we die? Why don't we see God? Why did Jesus come to earth? Only the Book of Scripture has answers to these questions. The Book of Nature does not—it cannot even explain its own assumptions.

The ninth question is: can the Two Books ever be in conflict with each other? No, they shouldn't. The idea that the Two Books can never be in conflict with each other is essential here. This idea goes at least as far back as St. Thomas Aquinas. He in fact rejected the idea of what Islamic philosophers had called "double truth," which meant that a notion could be true in theology or religion and, at the same time, false in philosophy or science. Instead Aquinas asserted that what we know through reason can never be in conflict with what we know through faith, and what we know through faith can never be in violation of what we know through reasoning. These two sources of information, according to Aquinas, can never be in conflict with each other—as long as we understand them correctly. Let's translate that here as follows: there can be no conflict between what the Book of Scripture tells us and what the Book of Nature tells us. There is no "double truth."

Therefore, enforcing some kind of separation between the Two Books makes perfectly sense in order to promote mutual respect between science and religion, as good fences make good neighbors. It is in their respective domains that science and religion function best. Besides, separate domains have recognizable boundaries which prevent merging and intermixing, thus also barring the possibility of partial or complete absorption. Problems between science and religion tend to arise when boundaries are being neglected. If there are any apparent conflicts between science and religion, then they are born of either bad science or bad religion, and they should compel the puzzled thinker to dig deeper and think harder.

The only time the Church herself may interfere in science is when scientists overstep their boundaries and venture into moral, philosophical, or religious territory. The advice of Cardinal Cesare Baronius expressed at the time of Galileo remains priceless, "The Bible teaches us how to go to heaven, not how the heavens go." So reading the Book of Scripture does not tell us how the heavens go, nor does reading of the Book of Nature tell us how to go to Heaven. As it turns out, science in itself is incomplete, and religion on its own is insufficient. That's why we need both of them.

The tenth question is: can the Two Books ever "contaminate" each other? Undoubtedly, there have been situations and debates where scientific theories were defended or attacked on religious grounds, or where religious views were defended or attacked on scientific grounds. A case in point is the famous, or infamous, controversy around Galileo. In 1992, Pope John Paul II held a conference celebrating the 350th anniversary of the publication of Galileo's *Dialogue Concerning the Two Chief World Systems*, at which he remarked that the experience of the Galileo case had led the Church "to a more mature attitude and a more accurate grasp of the authority proper to her," enabling her better to distinguish between "essentials of the faith" and the "scientific systems of a given age."

One could argue that we have learned from these experiences to look differently at their "interface." Scientific issues were eventually settled by more and better data and by considerations that were purely "scientific" in the modern sense. And religious issues were eventually settled by learning to define more properly what is essential to the religious faith. At the time of Galileo, for example, both sides were in a process of shaping their own identity. In other words, it became more and more clear what science is, what religion is, and what the difference is between these two. That process required some time and reflection.

Nowadays we know the outcome, but that wasn't quite clear at the onset. This in itself is not quite unusual. Just think of the way astronomy had to distinguish itself from astrology, at the time, and how chemistry had to distance itself from alchemy. They all acquired more distinctiveness during the process. What we should learn from this, though, is to never read the Book of Nature as if it were the Book of Scripture, nor vice-versa.

Based on this important caveat, we should read both Books in their own setting. On the one hand, the Book of Nature cannot deny or reject God on scientific grounds. On the other hand, neither can the Book of Scripture infiltrate science with a "god-hypothesis" to fill in the gaps still left behind by science. What this latter basically amounts to is making the First Cause occasionally act like a secondary cause working inside the Universe. This changes God into a "god-of-the-gaps"—a god who fills in the gaps left behind in his own allegedly defective design of the Universe. This kind of god often proves to be a fleeting illusion, for when the frontiers of science are being pushed back—and they usually are—this kind of god would be pushed back with them as well. The great scholastic theologian Francisco Suárez put it this way: "God does not intervene directly in the natural order where secondary causes suffice to produce the intended effect." That's where we need good fences.

Let's come to a conclusion. The distinction between the Book of Scripture and the Book of Nature may be rather ancient, but it still proves very useful in modern times. The separation of the Two Books makes clear that science addresses issues that are off-limits for religion, and religion addresses issues that are out-of-bounds for science. Keeping them distinguished prevents trespassing attempts and border conflicts. Again, good fences make good neighbors.

In general it is wise to avoid potential conflicts by respecting the fences between them. Therefore, science should not invade the domain of religious issues, and religion should not intrude into the domain of scientific issues. Science and religion are not competitors; they are rather neighbors, each one with their own domain, perspective, and authority. So each offers a one-sided story, but cannot deny there is another side to the story. That's why there are Two Books, instead of one. The "town" we live in is definitely big enough for both of them—neither one has to leave.

18. God and the Big Bang

In this chapter, we will see how the image of the Two Books may be helpful to better understand what the Big Bang (found in the Book of Nature) has to do with God (found in the Book of Scripture), and vice-versa. The image of the Two Books helps us explain why there is no conflict here, no either-or issue— which is contrary to what several partisans have argued. Some extremists from the religious camp have claimed that the Big Bang theory cannot be true if God is the Creator of the Universe. Other extremists, this time from the scientific camp, have claimed that there is no longer any need for a Creator if the Big Bang theory is true. Why are both factions wrong on this issue?

Here is what the Book of Nature tells us. Our Universe most likely started with the Big Bang, some fourteen billion years ago. Edwin Hubble had discovered in 1929 that the distances to far away galaxies were generally proportional to their red-shifts. Red-shift is a term used to describe situations when an astronomical object is observed to being moving away from the observer, such that emission or absorption features in the object's spectrum are observed to have shifted toward longer wavelengths—red, that is. Hubble's observation was taken to indicate that all very distant galaxies and clusters have an apparent velocity directly away from our vantage point—the farther away, the higher their apparent velocity.

This phenomenon had already been suggested in 1927 by the Belgian priest, astronomer, and physicist George Lemaître of the Catholic University of Louvain. In 1931, Lemaître went even further and suggested that the evident expansion of the Universe, if projected back in time, meant that the further in the past the smaller the Universe was, until at some finite time in the past, all the mass of the Universe was concentrated into a single point—a "primeval atom," in his own words—where and when the fabric of time and space started.

The English astronomer and mathematician Fred Hoyle is credited with coining the term *Big Bang* during a 1949 BBC radio broadcast. Currently, the Big Bang theory is the prevailing cosmological model that explains the early development of the Universe. According to this theory, the Universe was once in an extremely hot and dense state which expanded rapidly. This rapid expansion caused the Universe to cool and resulted in its present continuously expanding state—not with galaxies moving through space, but rather the space between the galaxies stretching.

Once it had cooled sufficiently, its energy was allowed to be converted into various subatomic particles, including protons, neutrons, and electrons. Giant clouds of these primordial elements would then coalesce through gravity to form stars and galaxies, and the heavier elements would be synthesized within stars and then be released during the explosion of supernovas. Interestingly enough, the 92 elements we find on Planet Earth can be found all over the Universe, indicating a common origin. They are produced very gradually inside the "furnace" of a star. So the presence of certain elements gives us an indication about the age of planets and stars.

There is little doubt among cosmologists that we live in the aftermath of a giant explosion that occurred some fourteen billion years ago. The first estimates of the age of the Universe based on its rate of expansion gave about a billion years, which was inconsistent with the known ages of the oldest rocks and of stars. But those first estimates turned out to be based on a mistaken calculation of the distances between galaxies. Eventually things got sorted out, and evidence for the Big Bang built up to the point where it is regarded as rather conclusive. Although nothing in science is final (§7), cosmologists will tell us that the Big Bang theory is still the latest and best we have; it remains standing for now until further notice. That's the reading of the Book of Nature.

But there is more to it. The introduction of the Big Bang theory has not only made quite an impact on cosmology, but also on philosophy and theology. If we have to believe some of the leading physicists and astrophysicists nowadays, then the Big Bang story is the modern replacement of the "old" creation story, leaving no longer room for a Creator of this Universe. There are several ways they have come up with the idea that "creation out of nothing" [*creatio ex nihilo*] is no longer a religious or philosophical concept that requires a Creator.

The British cosmologist Stephen Hawking, for instance, talks about the Big Bang in terms of what he calls a "spontaneous" creation: "Because there is a law such as gravity, the Universe can and will create itself from nothing. Spontaneous creation is the reason there is something rather than nothing." This in turn made the late astrophysicist Carl Sagan exclaim, in the preface of one of Hawking's books, that such a cosmological model has "left nothing for a creator to do." Others, such as the cosmologist Lee Smolin, also made sure there is no space left for a Creator by proclaiming that "by definition the Universe is all there is, and there can be nothing outside it."

Another attempt to eliminate the idea of "creation out of nothing" was made by Alexander Vilenkin. He developed an explanation of the Big Bang in terms of "quantum tunneling from nothing"—as a fluctuation of a primal vacuum. Just as sub-atomic particles appear to emerge spontaneously in vacuums in laboratories, as a result of what some have called "quantum tunneling from nothing," so the whole Universe may be the result of a similar process, he believes. Again, if true, that would be the end of any belief in a Creator. Stephen Hawking, for instance, doesn't get tired of declaring that science provides a "more convincing explanation" for the Universe than God. Where do these cosmologists go wrong?

It may seem at first sight that this is the end of religion, creation, God, and the Book of Scripture. However, there is quite some confusion underlying this debate. First we have to discuss what the idea of "creation out of nothing" [*creatio ex nihilo*] really stands for, before anyone can wipe that idea from the table. St. Thomas Aquinas can help us disentangle some knots. He would say that when philosophers speak of "creation out of nothing," they certainly are not talking science, let alone physical cosmology. They are talking in terms of the First Cause, who is not a secondary cause, and thus not an object of science.

Aquinas also makes an important distinction between the

concept of producing [*facere*] and the concept of creating [*creare*]. Science is about "producing" something from something else—it is about secondary causes, about changes in this Universe, about changes from one thing to another thing, about "like causes having like effects." Science is about these secondary causes, which necessarily leaves the First Cause untouched.

Creation, on the other hand, is about "creating" something from *nothing*—which is not a change at all; certainly not a change from "nothing" to "something." In other words, the Creator doesn't just take pre-existing stuff and fashion it, as does the Demiurge in Plato's *Timaeus*. Nor does he use some kind of something called "nothing" and then create the Universe out of that. Rather, God as First Cause calls the Universe into existence without using pre-existing space, matter, time, or whatever. Whereas science deals with secondary causes, creation has everything to do with the First Cause.

In other words, creation is not a change. Instead, it's a cause, but of a cause of a very different, indeed unique, kind. Therefore, *creating* something "out of nothing" is not *producing* something out of nothing—which would be a conceptual mistake, for it treats nothing as a something, and it confuses "creating" with "producing." On the contrary, the Christian doctrine of creation *ex nihilo* claims that God creates the Universe out of nothing—which is very different from making it out of something else. Creation has everything to do with the philosophical and theological question as to why things exist at all, before they can even undergo any change. Therefore, creation—but not the Big Bang—is the reason why there is something rather than nothing (including something such as the law of gravity). Consequently, "creation out of nothing" is a concept that does not belong to the Book of Nature but is an important part of the Book of Scripture. It has everything to do with the First Cause.

Seen in this light, Hawking's idea of a "spontaneous creation" is sheer philosophical magic, actually nonsense. Hawking is a good scientist, but that does not necessarily make him a good philosopher. For something to create itself, it would have to exist before it came into existence—which is logically and philosophically impossible. How could the Universe "create itself" from nothing—let alone cause itself?

To use a simple analogy, no son can be his own father. The law of gravity cannot do the trick, for before the Universe could ever supposedly "create itself" through the law of gravity, we have to posit the existence of laws of physics—which are ultimately the laws which govern the existing Universe. So Hawking is actually saying that laws which have meaning only in the context of an existing Universe can generate that Universe (and its laws of nature) before either exists—which makes, again, for a logical contradiction. Even in mathematics, we cannot prove anything from nothing; at least something is needed, axioms, to start the process.

St. Thomas Aquinas would join the great philosopher Aristotle in responding that whenever there is a change, there must be something that changes, for nothing comes from nothing. Plenty of nothing is still nothing—you can have "plenty" of it, but it is still nothing. All change requires some underlying material reality. So it is a mistake—a category-mistake if you will—to use arguments coming from the natural sciences to deny the concept of creation in philosophy and theology. The theologian David Bentley Hart puts it this way: "Physical reality cannot account for its own existence for the simple reason that nature—the physical—is that which by definition already exists." Many scientists fail to see this point. Perhaps Albert Einstein was right after all when we quoted him in the preface, "that the man of science is a poor philosopher."

Apparently, the real nature of the concept of "creation out of nothing" has eluded many cosmologists. Hawking, as we saw, tells us that gravity would be able to "create" the Universe—spontaneously, so to speak, or automagically. And the British physical chemist Peter Atkins claims that science has a limitless power and must even be able to account for the "emergence of everything from absolutely nothing." In response to such jumbled claims, St. Thomas Aquinas would keep hammering on the distinction between producing and creating, or between changing and creating, or between secondary causes and First Cause.

God, being the First Cause, operates in and through secondary causes, including the Big Bang. But the Big Bang does not bring anything into "being"; it works with something that is already in existence, thanks to creation. We cannot put these two concepts on the same level, let alone in competition with each other. Creation is very different from what science is searching for. Science can only investigate what already exists, but creation is the source or cause of all that exists. We have here a vital distinction that helps us avoid a jumbled discussion.

Not only is there a misunderstanding in the debate about the meaning of "creation," but also about the meaning of "nothing." It needs to be stressed that "nothingness" in the expression "creation out of nothing" is not a highly unusual kind of exotic "stuff" that is more difficult to observe or measure than other things; it is not some kind of element that has not found a position yet in the periodic system; it is in no way a material thing that can change into something else, but it is actually the *absence* of anything—and therefore we cannot treat no-thing as a some-thing. Therefore, "creation out of nothing" does not mean changing a no-thing into a some-thing, or changing something into something else—like chemists change water into hydrogen and oxygen. Instead it means bringing everything into being and existence. Again, that point deludes many scientists. They don't realize that "nothing" in physics is really "some-thing," whereas in metaphysics it refers literally to "no-thing."

When Alexander Vilenkin tries to explain the Big Bang in terms of "quantum tunneling from nothing"—as a fluctuation of a primal vacuum—he falls into that same trap. True, when an electron and a positron collide, they can "an-nihil-.ate" and thus change into "nothing" [*nihil*]. What really happens, though, is that when they annihilate, they emit a burst of energetic photons—which is certainly not "nothing" in a metaphysical sense. On the other hand, the reverse can occur too, when an "empty" space is filled with an electric field, but without particles. In that situation, there is a certain probability that suddenly an electron-positron pair will pop out of "empty" space.

In other words, "quantum tunneling," causes some "system" to change from one "state" (an electric field without particles) into another "state" (a changed electric field with two particles). These are different "states," but of the same "system." However, it must be realized that a pair of particles does not suddenly appear out of literally "nothing," but actually out of an electric field of an existing "system"—which, again, is not nothing in a metaphysical sense.

So it makes no sense to use quantum tunneling at the time of the Big Bang to make a Universe suddenly pop into existing from "nothing" by a quantum fluctuation. As William E. Carroll puts it succinctly, "it is still something—how else could 'it' fluctuate?" We are dealing here again with a "system" that has a set of possible "states." Obviously a state is a state, certainly not "nothing." It is a specific "state" of a specific, complicated quantum "system" governed by definite laws. Stephen Barr uses the analogy of having a bank account with no money in it: even if we have "nothing" in the bank, we still have a bank, with all that comes with it, but it happens to be in a "no-money state" for us. This kind of "nothing" is different from having no bank or bank account at all.

Therefore, we need to recognize that the "nothing" discussed by present-day cosmologists is not absolutely nothing in the literal, metaphysical sense. "Nothing" does not exist in the Book of Nature—it belongs to the Book of Scripture. Those who claim that a "vacuum" or "nothing" can be described mathematically with a wave function—which they consider the quantum gravity equivalent of the quantum vacuum in quantum field theory—seem to entirely miss the point that such a wave function is definitely something rather than nothing. In other words, scientists sometimes use the same terms as philosophers and theologians, but they use them in a very different sense, making them different concepts. So what scientists call a "vacuum" or "annihilation" or "nothing" is very different from the philosophical concept of "nothing" (*nihil*) in "creation out of nothing."

Put differently, a "vacuum" in physics is not "nothing," and "nothing" in philosophy is not a "vacuum"; these two terms are not the same but belong to two very different vocabularies from two different Books, and thus make for different concepts. Actually, even with the most modern techniques, we have never been able to pump all atoms and molecules away from a so-called vacuum. Scientists are getting better at it, but never has a vacuum become literally "empty." Since matter is basically a form of energy, there is always some matter or energy left in a vacuum. Nowadays, it is even postulated that a vacuum is filled with an infinite number of electrons and positrons (with the same mass as electrons but with a positive electric charge). Again, a "vacuum" in physics is different from "nothing" in philosophy and theology.

Nevertheless, some scientists maintain that the origin and existence of the Universe is an entirely physical issue. If there is any metaphysical assumption lurking behind their erroneous claim, it is that the mere existence of things needs no further explanation. This view has rightly been caricaturized by the Boston College philosopher Peter Kreeft as a magical "pop theory" that has things pop into existence without any cause. A Higgs boson, for instance, cannot just pop itself into existence; it must have a cause, because it does not and cannot have the power to make itself exist, for then it would have to exist before it came into existence. This may not be a logical necessity, but it is a metaphysical necessity, for it is impossible for things to pop into being without a cause—"out of nothing," that is. In science, "nothing" is a concept very different from "nothing" in religion. Missing this distinction basically amounts to denying the existence of Two Books, instead of one.

This takes us to another sort of confusion in this debate: Was creation the beginning of the Universe? Let us make clear first that there is no logical contradiction in the notion of an eternal Universe. An eternal Universe would be no less dependent upon God, the First Cause, than a Universe that has a beginning of time. Even if there had always been a Universe and that it never "began," even then could it only exist at any moment in time because the First Cause is causing it to and keeping it in existence. It could be possible, for instance, that explosions and collapses follow each other in an endless sequence of expanding and contracting. If so, the Universe might have never had a beginning. Aristotle, for one, believed in a first cause, but not in a beginning of the Universe. But in 1215, the Fourth Lateran Council taught that the Universe was created "out of nothing at the beginning of time"—an idea which would have scandalized both ancient Greeks and nineteenth-century positivists, but which is now a common-place of modern cosmology.

Yet, if there was a beginning—which is affirmed by the Book of Scripture and is nowadays also accepted by the Big Bang theory coming from the Book of Nature—then we have reason to ask: was the Big Bang "preceded" by creation; was it creation that triggered the Big Bang; was creation the event that happened "before" the Big Bang took place, or could even take place? These seem to be legitimate questions, but only if we take "before" in the sense of "causally prior," not "temporarily prior." In that specific sense, creation must come "first" in the order of causality before any events, even a Big Bang, can follow in the order of time.

However, creation is not an event at all. Creation concerns the *origin* of the Universe—its source of being—not it's *beginning* in time. The Book of Nature can deal with the beginning of time ("*at* the beginning"), but the Book of Scripture deals with the origin of time and the Universe ("*in* the beginning"). We have two different concepts of "beginning" here. Creation does not have a beginning or an end, but the Universe does.

This means creation is not some distant event in the Book of Scripture. Instead it is the complete causing of the existence of everything that is—in the past, now, and in the future. Creation must come "first" in the order of primacy, not in the order of time. The creation of the First Cause is not first in a temporary sense but in an explanatory sense; it is about causal priority, not temporal priority. Creation itself is just not another cause in a temporal sequence of secondary causes, but it is linked to the First Cause which makes all secondary causes possible. Put more technically, creation is not temporarily prior to the Big Bang, but it is causally prior.

Those who claim instead that certain things have the

capacity to continue existence on their own—without God, that is—must face the problem that this very capacity remains unexplained without God. To explain the existence of things, including the continuation of their existence, requires creation at every moment in time. All material things depend on the First Cause for their existence as secondary causes—at every moment of their existence. Without a First Cause, secondary causes could not exist.

Yet, it is thanks to the Big Bang Theory that we can now raise the question of the beginning of time. Obviously, the unfolding of the Universe, starting with the Big Bang, is a process that plays in time and can be studied by the Book of Nature. Creation, on the other hand, cannot follow a timeline, as time itself is a product of creation as well. Time is something that *began* at one point—that is, time had to come into existence at one point. Albert Einstein had already showed us that both time and space are part of the physical world, just as much as matter and energy. In point of fact, time can be manipulated in the laboratory. The presence of mass (and more generally energy) causes space-time to curve. And near the horizon of a black hole, where the curvature of space-time is quite infinite, black holes stretch time by an infinite amount—that is, black holes cause infinite "dilation of time."

If time is indeed part of the physical world, then we cannot place creation at the beginning of time, since there is no time until time has been created. Creating time "at a certain time" is just tough to do! God himself is time-less. Therefore, creation is not something that happened long ago in time, and neither is the Creator someone who did something in the distant past. Instead, the Creator does something at all times—by keeping a contingent world in existence. Whereas the Universe may have a beginning and a timeline, creation itself does not have a beginning or a timeline; creation actually makes the beginning of the Universe and its timeline possible.

Put differently, creation creates chronology, but it is not a part of chronology. If time started with the Big Bang, then it does not make sense to ask what happened "before" the Big Bang, because there was no time yet until time had been created. We should never confuse temporal *beginnings* in the Book of Nature with metaphysical *origins* in the Book of Scripture. Once we lose sight of this important distinction, we are in for a dangerous mix-up. The image of the Two Books can save us for that error.

If rightly understood, creation is not a "one-time deal," but instead it copes with the question as to where this Universe ultimately comes from, as to how it came into being and how it stays in existence. The answer is that it does not come from the Big Bang, but may have started with the Big Bang. Without creation, there could not be anything—no Big Bang, no gravity, no evolution, not even a timeline. Creation sets the "stage" for all of these things and keeps this world in existence. It explains why there is something rather than nothing. Hence, we cannot put the "seven days of creation" in the first chapter of the Book of Scripture against the scenario of the Big Bang in the Book of Nature. In other words, the Book of Scripture does not have a scientific theory of the world's beginning, but rather a religious creed about the world's origin and foundation.

The "rest of the story" would be something for science to tell—and science is definitely trying hard to do so. Take the analogy of a novel: the *beginning* of the novel consists of its first words or sentences; but the *origin* of the novel is what the author of the novel has come up with. This allows Carroll to state, "We do not get closer to creation by getting closer to the Big Bang," for the simple reason that creation is not an event in the explanatory domain of cosmology—it is a metaphysical and theological concept regarding the *origin* of the Universe. The

Big Bang did not "create" time, any more than it "created" gravitation, let alone the law of gravity. Only creation *ex nihilo* does! Let's not forget there are Two Books to deal with, not just one. Creation is something "in the beginning," which made it possible for something to happen "at the beginning." In science, "beginning" is a concept very different from the concept of "beginning" in religion.

It must be clear by now that science and religion cannot discard each other. This statement should be "breaking news" for all those who see a conflict between creation, coming from the Book of Scripture, and the Big Bang, coming from the Book of Nature. What we have been trying to prove in this book is that science and religion need each other. On the one hand, science cannot discard religion, because religion has concepts from the Book of Scripture that science doesn't and can't know about. In short, science cannot discard religion.

On the other hand, religion cannot reject what science tells us through the Book of Nature either. St. Thomas Aquinas could not have said it more clearly, "The truth of our faith becomes a matter of ridicule among the infidels if any Catholic, not gifted with the necessary scientific learning, presents as dogma what scientific scrutiny shows to be false." Religion cannot ignore science. Hence, neither science nor religion have to leave "town"—there is definitely space for both of them. The "town" we live in needs them both.

19. God and Evolution

The image of the Two Books is also very helpful to take on the seeming conflict between God and evolution. Where do human beings come from? I am not saying you have to accept the theory of evolution if you are Catholic; evolutionary theory is acceptable for Catholics, but it's certainly not a religious dogma. But if you do accept this theory, would that mean you are in for a clash with your religion? A looming conflict seems to be almost inevitable: if we came from God, then we did not emerge from evolution; if we came from evolution, then God has nothing to do with it. That seems to be a stalemate, until we bring in the image of the Two Books again. Let's see what we can learn from that image.

A few centuries ago, James Ussher (1581-1656), the Protestant Archbishop of Armagh in the Church of Ireland, calculated a chronology of the history of the world according to the Book of Scripture. Based on a literal reading of the Old Testament, Ussher deduced that the first day of creation began at nightfall on Saturday, October 22, 4004 BC. If Ussher could be that precise, so many think, then his "date of creation" must have been very reliable and accurate. No wonder then that his calculations, or similar ones, are still very popular in some Protestant circles that reject the centuries-old distinction between the Book of Scripture and the Book of Nature.

What is wrong with Ussher's chronology? The problem is not so much that the exact date might be off by a certain number of years, but rather that Ussher was reading the Book of Scripture as if it were the Book of Nature. We said earlier (§17) that the chronological interpretation of the first chapter of Genesis is very questionable. But what is more important, the Book of Scripture is not a college text book on physics, chemistry, or biology. It is not about science—there was hardly any science at the time it was written—but it is about God, about the Creator of this Universe, about the origin and destination of humanity. Reading it literally and with a "scientific" mindset does great injustice to the Book of Scripture. It turns religion into a semi-science.

How then should we read the message of the first chapter in

the Book of Scripture? We mentioned already (§17) that this chapter has two sections: the work of division, which divides things from each other, and the work of adornment, which populates the various realms. It is through this structure that Genesis 1 proclaims its core message of *monotheism* against the pagan beliefs of polytheism that surrounded Israel. It emphatically proclaims that nothing exists that does not owe its existence to God the Creator. The Book of Scripture does not teach us science but belief in one God, the Maker and Master of Heaven and Earth. Its core message is sound monotheism.

This Biblical message was, and still is, unique. Most of the peoples surrounding Israel regarded the various regions of nature as actually *divine*: they worshiped the sun, moon, and stars as gods; they had sky gods, earth gods, water gods, gods of light and darkness, of rivers and vegetation, of animals and fertility. To battle this form of polytheism, Genesis 1 proclaims its radical and imperative affirmation of monotheism. Each day dismisses an additional cluster of deities: on the first day, the gods of light and darkness; on the second day, the gods of sky and sea; on the third day, earth gods and gods of vegetation; on the fourth day, sun, moon, and star gods (including astrology); on the fifth and sixth days, gods from the animal kingdom (such as sacred falcons, lions, serpents, and golden calves). Finally, even humans are emptied of any *intrinsic* divinity, even while they are *granted* a divine likeness.

So each "day of creation" shows us how another set of idols is being smashed. Nothing on Earth is god, but everything comes from God. Obviously, the issue at stake in Genesis 1 is monotheism, not science—not physics or biology, let alone evolution. We are dealing here with the Book of Scripture, not the Book of Nature.

The Book of Nature tells us an entirely different story. Put in a nutshell, astrophysics estimates the Big Bang happened at approximately 13.8 billion years ago, which is thus considered the age of the universe. Astronomy tells us the age of the Earth is 4.5 billion years, based on evidence from radiometric age dating of meteorite material. Geology adds to this that ancient rocks exceeding 3.5 billion years in age are found on all of Earth's continents. (Earth's oldest rocks were recycled and destroyed by the process of plate tectonics.) And paleontology shows us that fossils range enormously in age, and can be up to 3.48 billion years old, or even as old as 4.1 billion years.

The timescales of the Book of Nature are obviously very different from the timescales used in the Book of Scripture, not to mention Ussher's chronology. The time spans we are dealing with in science are so huge that they defy human imagination. We cannot really fathom how immense these numbers are. What may help us to get a better "feel" of such enormous magnitudes is using the timeline of a 24-hour *day*, with the Big Bang taking place at the beginning of the day, at 12:01 AM, and the current time being at the end of the day, at 11:59 PM.

On this scale, each hour of the day would represent some 600 million years. That's still beyond human imagination, but at least it makes it easier for us to compare how long it took for certain events to take place. On this scale of a day, Planet Earth would begin to form at 4:17 PM, and life would begin to develop at 5:08 PM. Seen this way, humanity appeared on Earth at the very last minute—not sooner than 11:59 PM. It was a "last-minute" step, some say—perfectly in time, or perhaps almost too late.

Using the timeline of one calendar *year*, instead of a day, would amount to ±38 million years per day. After the Big Bang had taken place at the beginning of January, planet Earth would then start to form at the beginning of the month of September. Then life would emerge on Earth during the middle of the month of September, Dinosaurs would appear around

December 24, and mammals around December 28. Then, even closer to the end of the year, human beings would emerge at about December 29.

This raises at least two serious questions. Here is question one: why does the Universe have to be so *old*? Well, to put it briefly, chemicals had to be formed in stars and then be released when they exploded as supernovas. The whole lifetime of a star is typically billions of years. So living creatures cannot appear until at least some stars have had time to explode and release the elements needed for life's chemistry. And then there is question two: why does the Universe have to be so *big*? Because of its expansion, the size of the Universe is directly related to its age. It has grown to a size of at least 15 billion years. If the Universe were any smaller, it would not have lasted long enough for life to emerge.

But behind all of this is a more challenging question: how do we know all these "dates," which are extremely different from what Ussher thought? How do we know the age of the Universe, for instance, given the fact that no one was there? There are several ways scientists have come to an approximate age of 13.8 billion years. One of the measurements used to calculate the approximate age of the Universe is based on the expansion rate of the Universe. The universe does not expand "into" anything, but space itself is expanding. Scientists can use the Hubble Space Telescope to measure the rate of this expansion. Based on this rate, they can extrapolate backwards in time—to the point where all the mass of the Universe was concentrated in a single point, which was the event of the Big Bang.

So if we know the expansion rate of the Universe, we have a better idea of when the Universe was born. The first reasonably accurate measurement of the rate of expansion of the Universe—a numerical value now known as the Hubble constant—was made in 1958 by astronomer Allan Sandage. His measured value for the Hubble constant came very close to the value range generally accepted today. But in science, nothing is etched in stone. Yet, it's the best we have right now.

That's how we know quite precisely when the Universe was born. But what about the Earth? How do we know when the Earth was born and when the Earth would hold its first fossils as an indication of life? Fortunately we have much better scientific ways of dating nowadays than a few decades ago. The most important technique is radioactive dating. Here follows a very basic explanation of isotopes first.

On Planet Earth we know of 92 different, naturally occurring chemical elements, ranging from atomic number 1 (hydrogen) up to 92 (uranium). Elements with higher atomic numbers have only been synthesized in laboratories or nuclear reactors. The so-called Periodic Table of Elements shows all of them in a specific order, based on their atomic number. The number of protons determines the atomic number (1 to 92) and defines what element it is and also determines the chemical behavior of the element. For example, hydrogen atoms have one proton, carbon atoms have six, and oxygen atoms eight.

Every chemical element has its own kind of atomic nucleus, made up of a specific number of protons—their atomic number—together with some number of neutrons, the latter of which is different for different "isotopes" of the element. An atom of the element carbon, for instance, has exactly six protons in its nucleus. One isotope of carbon has also six neutrons, making for a total of twelve (6+6=12). Therefore it is called "carbon 12." Another carbon isotope has eight neutrons, which makes for "carbon 14" (6+8=14). The simplest and smallest element is hydrogen, with one proton and one neutron in its nucleus, "hydrogen 2."

How can these elements help us to date things on Earth?

The secret can be found in the existence of radioactive isotopes. As said before, isotopes of a particular element differ in the number of neutrons in their nucleus. Some isotopes are inherently unstable. That is, at some point in time, an atom of such an isotope will undergo radioactive decay and spontaneously transform into a different isotope. Radioactive isotopes have a known and constant rate of radioactive decay. These isotopes always start with a known level of radioactivity at the time they cooled from molten lava and became solid.

While the moment in time at which a particular atomic nucleus decays is unpredictable, a collection of atoms of a radioactive isotope decays exponentially at a rate described by a parameter known as the half-life, usually given in units of years when it comes to dating techniques. After one half-life has elapsed, one half of the atoms of the isotope in question will have decayed into a "daughter" isotope or decay product. In many cases, the daughter isotope itself is radioactive, resulting in a decay chain, eventually ending with the formation of a stable (non-radioactive) daughter isotope.

For most radioactive isotopes, the half-life depends solely on nuclear properties and is essentially a constant. It is not affected by external factors such as temperature, pressure, chemical environment, or presence of a magnetic or electric field. So, the proportion of the original isotope to its decay products changes in a predictable way as the original isotope decays over time. This predictability allows the relative abundances of related isotopes to be used as a "clock" to measure the time from the incorporation of the original isotopes into a material to the present. Take, for instance, radioactive radium-226 (with 88 protons and 138 neutrons). By emitting a so-called alpha particle, which has 2 protons and 2 neutrons, this isotope decays into radon-222 (with 86 protons and 136 neutrons). Even though the half-life of this radioactive decay can be determined (1,600 years), nothing determines when one particular atom will disintegrate. Nevertheless, radioactive decay occurs at a regular pace and thus can be used as a "clock."

To summarize, radioactive dating is now the principal source of information about the absolute age of rocks and other geological features, including the age of fossilized life forms as well as the age of Planet Earth itself. So, thanks to the Book of Nature, we have come a long way in the right direction since the time James Ussher made his calculations.

It is time now to go back to our original question: can creation and evolution live together? What has God to do with evolution? I assume for now that you accept there is/was evolution on Planet Earth. How it happened—its mechanism— is an issue I won't go into here. I will only focus on the question as to whether "belief" in evolution is compatible with belief in God. To answer this question, it might be helpful to see what St. Thomas Aquinas would have said about this issue, although the modern concept of evolution was unknown to him.

You don't have to believe in evolution to know that human beings are part of the animal world, seen from a biological point of view, because we share many characteristics with the animal world—we all breed, feed, bleed, and excrete. Even St. Thomas Aquinas knew this, more than six decades before Charles Darwin. He called human beings animals, but set them apart as rational animals [*animal rationale* or *rationabile*]. Aquinas did not need any learned biology to see the obvious truth that humans are first of all animals.

All mammals, for instance, share basically the same morphological structure, in spite of obvious differences in appearance. Just the fact that all of them have seven cervical vertebrae illustrates this point—no matter whether it is in the sturdy neck of a rhino or the long neck of a giraffe. And the same holds for human beings. So there are similarities and dissimilarities in the animal world. Aquinas even knew that "animals of new kinds arise occasionally from the connection of individuals belonging to different species, as the mule is the offspring of an ass and a mare."

Obviously, Thomas Aquinas did not know about evolution the way we understand that concept today. Although a few scholars maintain that Aquinas would not have accepted natural evolution, it could be argued he would probably have said that evolution offers us a scientific account of changes, of how a later state of the material world might have emerged from an earlier state by the action of secondary causes during the process of evolution. Creation, on the other hand, offers us a metaphysical account of where the material world itself ultimately comes from (§5). In this line of thought, creation is about the First Cause, whereas evolution would be about secondary causes such as mutation, speciation, and natural selection.

In line with what we said earlier (§18), creation *creates* something out of nothing, whereas evolution *produces* something out of something else—that is, by following biological laws in the same way as planetary motions follow physical laws. God does not make things himself—in a manual and interventional way, so to speak—but God makes sure they are being made through his own laws of nature and secondary causes. God as the First Cause operates in and through secondary causes, including the causes of evolution. That would probably be Aquinas' take, if I may say so.

Interestingly enough, even Charles Darwin explained the natural evolution of species as due to what he literally called "secondary causes." In distinguishing "secondary causes" from a "First cause," Darwin adopted a metaphysical principle that was already formulated by St. Thomas Aquinas. This allowed

Darwin to say, "To my mind it accords better with what we know of the laws impressed on matter by the Creator, that the production and extinction of the past and present inhabitants of the world should have been due to secondary causes, like those determining the birth and death of the individual." Kudos to Darwin, but first of all to Aquinas!

Thanks to the philosophy of St. Thomas Aquinas, the Catholic Church does not have any fundamental problems with the issue of "God AND evolution." She also honors the distinction between the Book of Nature and the Book of Scripture. Because of this position, the Catholic Church has kept an open mind regarding the biology of evolution. As early as 1950, Pope Pius XII wrote in his encyclical *Humani generis*, "The magisterium of the Church does not forbid that, in conformity with the present state of human sciences and sacred theology, research and discussions, on the part of men experienced in both fields, take place with regard to the doctrine of evolution, in as far as it inquires into the origin of the human body as coming from pre-existent and living matter." But he also added emphatically, "the Catholic faith obliges us to hold that souls are immediately created by God."

In April 1985, Pope John Paul II gave an address to a symposium on evolution, in which he said, "Rightly comprehended, faith in creation or a correctly understood teaching of evolution does not create obstacles: Evolution in fact presupposes creation; creation situates itself in the light of evolution as an event which extends itself through time—as a continual creation—in which God becomes visible to the eyes of the believers as 'Creator of heaven and earth.'"

Then, in a 1996 Message to the *Pontifical Academy of Sciences*, Pope John Paul added, "Today, [...] some new findings lead us toward the recognition of evolution as more than a hypothesis. It is indeed remarkable that this theory has been progressively accepted by researchers, following a series of discoveries in various fields of knowledge." His successor, Pope Benedict XVI, spoke along the same lines. On July 26, 2007, he rejected the idea that "whoever believes in the creator could not believe in evolution, and whoever asserts belief in evolution would have to disbelieve in God. [...] This contrast is an absurdity, because there are many scientific tests in favor of evolution." Then he added emphatically, "But the doctrine of evolution does not answer all questions, and it does not answer above all the great philosophical question: From where does everything come?"

Apparently, in all the previous statements, the Catholic Church is very aware of her own authority. She cannot endorse evolutionary theory, heliocentrism, or the Big Bang theory, but neither can she reject them—for either verdict would be beyond her competence. She realizes she cannot use the Book of Scripture to answer questions about the Book of Nature. We always need to find out whether a certain scientific theory is true or not. If it is true, religion will have to accept it, given the fact that religion seeks the truth. If it is not true, present-day science will have to modify itself, since science seeks the truth as well.

Not so long ago, most Catholics in the pews understood that evolution was somehow consistent with Church teaching. What has changed since is not Church teaching, nor evolutionary theory, but the fact that non-Catholic Fundamentalists and Evangelicals, especially in the USA, now have an enormous impact on our culture, as has their rejection of evolution, thus making Catholics feel they must be suspicious of evolution if they want to be faithful to their religion. So the issue of evolution has become highly suspect, even with some Catholics—no matter what Roman pontiffs and prominent Church theologians and cardinals have said to the contrary.

What can we learn from this? Against those who want to take the theory of evolution out of biology text books, I would say we should teach in our schools not only the theory of evolution but, if possible, also the philosophy of St. Thomas Aquinas, thus giving students not half but the full truth—thus keeping in mind the words of Pope John Paul II, "Evolution in fact presupposes creation." Only then can science and religion live together and enrich each other.

VII

Conclusion

First we discussed what science can do for us. The idea that science starts with "clean" observations was criticized. We found out that "pure" and "clean" observations do not exist—they are almost always "concept-laden" or "theory-laden." In other words, research begins with asking the right questions—by means of new concepts, models, hypotheses, and theories. But this step remains wishful thinking if not followed by testing their implications in the field or in the lab.

Seen this way, scientific research may be described as a "dialogue" between scientists and nature, between *possible* observations (hypotheses) and *actual* observations (facts). This process is done in what's commonly called the *empirical cycle*. In a nutshell, scientific research is a process leading from hypotheses to test implications, which in turn lead to observations that either confirm or falsify the hypotheses. This makes for a tentative enterprise, leading to acceptance, revision, or rejection of a hypothesis or theory, or even of an entire paradigm or research program. What is accepted today may be modified or even discarded tomorrow. Because the empirical cycle is a never-ending, cyclical process, science never ends.

This may easily lead to the idea that the scientific method is actually the best, even only, method we have to gain knowledge about this world—which amounts to an ideology called scientism. Scientism worships at the altar of science. We discussed many reasons why scientism cannot live up to its claims and promises. This criticism opened the door for other forms of knowledge, more in particular coming from religion. In short, science is not the only way of knowing.

Scientific research may well be called a fact-finding enterprise. Science is supposed to deal with facts, which makes for "scientific facts." Science has indeed come up with facts that are relevant in the light of scientific theories. However, not all facts are scientific facts. There are simple facts for which we don't need science—facts such as "Snow is white." There are also religious facts that go beyond the reach of science—facts such as "There is a God," or "God created the Universe." This means that most knowledge—not only in science but also in religion—is fact-based. So what we should expect from both science and religion, among other things, is that they provide us with facts. In other words, there are scientific facts as well as religious facts. Either set is composed of facts, but they differ in the way we know they are facts. For scientific facts, we need experimental tests and empirical observations. For religious facts, such tests are hardly ever possible. Yet, religious facts still need to be tested through reason and revelation as to whether they are real facts.

We also found out that the world of facts is a rather peculiar world. We analyzed why facts are not the clear-cut objects many people think they are. We came to the conclusion that a fact is the description of an event, the object of a thought, and the content of a statement, all at once. In other words, facts are interpretations of events by means of thoughts and statements. Through events, facts can be tested; through thoughts, they can be understood; and through statements, they can be communicated. It is through interpretation that thoughts and statements transform events into facts. But the bottom-line is this: unlike things, events, or situations, facts are non-material entities.

Not surprisingly, what facts have in common with observation statements is that they are not staring us in the face. They both require concepts. A further analysis revealed that concepts cannot be material entities located *outside* the mind, because material things are always particular, whereas concepts are always universal. Neither can concepts be material entities located *inside* the brain—for instance, neuronal firing patterns in the brain—for that would make them something particular too. Particular material things cannot qualify as universal. Besides, this explanation of concepts depends on other concepts such as neuron and brain—which creates a circular regression. Does this mean then that concepts are mere immaterial thoughts? If that were true, different minds would have different concepts, and there would be no way different people could communicate with each other—all of them would live in their own private, mental worlds.

To safeguard that concepts—and therefore also propositions which are built on concepts—exist as real objects, without being neither material nor mental, they would need to exist somewhere as abstract objects. This is not only true of concepts in science but also of concepts in religion. They need a realm for them to exist independently of the material world or of our thoughts about them. Concepts are like light beams that hit a certain thing that is in darkness, so this thing will be in the light and become visible.

These concepts exist outside our minds in a world of their own. So we came to the conclusion that concepts can only be explained as existing in the Divine Intellect, where they come from and reside—which is a conclusion that may be taken as another proof of God's existence. It is in the Divine Intellect that facts exist, scientific as well as religious facts. Facts are the content of propositions, and like all other propositions, they reside in the Divine intellect. That's also where their truth comes from.

After finding out what science can do—which is quite a bit—we had to discuss also what science cannot do—which turns out to be quite a bit too. We showed there are at least five assumptions science must accept in order to survive: the assumptions of reality, order, causality, comprehensibility, and testability. Science cannot even begin without those assumptions. However, what science cannot do is proving its own assumptions, yet they are needed for science to do what it can do. In other words, what science can do is only possible if it realizes what it cannot do—testing and proving its own assumptions, to begin with.

We actually discovered that these assumptions find their foundation in the Judeo-Christian religion. This means that scientists are in fact still living off Judeo-Christian capital, whether they realize it or not. This also explains why science could only come to life in cultures with a Judeo-Christian background. But once science had emerged on that hotbed, it gradually began to forget about its cradle. We showed how dangerous it would be to cut off the very roots from which science did spring up. Without God nothing has a firm foundation.

From this follows that putting science in conflict with religion is a no-no. They are not exclusive alternatives. To use a silly example, we don't have to decide whether the Grand Canyon was carved out by God or by the Colorado River. That's an absurd contrast. The Grand Canyon was made by river streams—secondary causes, that is—yet it was created by God—the First Cause—for without God there could be no river streams, actually no Grand Canyon at all. In a similar way it is absurd to say that children cannot come from God because they come from their parents, or that children cannot come from parents because they come from God. So to put science and religion in contrast to each other would be equally absurd.

Although science can certainly do a lot for us, there are at least two caveats. On the one hand, what science can do for us is of a limited scope—restricted to what can be dissected, measured, and counted. But science is not the only way of knowing. On the other hand, as we found out, science could never live without religion, because science works with assumptions that it could never validate on its own, but only with the help of religion—assumptions about reality, causality, intelligibility, order, and experimentation. In other words, science can only do a lot for us if it acknowledges that what it can do is only possible because of what it cannot do—testing and proving its own assumptions which have strong religious roots.

Once we have acquired a more realistic picture of science— of what science can and cannot do—and a more realistic picture of religion—of what religion can and cannot do—we should be able to see that science and religion can perfectly and harmoniously live together. This is very well expressed in the image of the Two Books—the Book of Nature and the Book of Scripture. Not only do we need to read the Book of Nature, which gives us more and more answers to our scientific questions, but also the Book of Scripture, which gives us answers to vital questions science could never deal with.

The image of the Two Books also helps us understand some seeming conflicts between science and religion. Those are conflicts based on dilemmas forced on us by people who do not understand science or do not understand religion. Famous examples are issues such as "God and/or the Big Bang" or "creation and/or evolution." A proper understanding of the power of religion and of the power of science tells us they are not contradictory to each other—they make for *and*-issues, not *or*-issues. The Book of Nature should never trump the Book of Scripture, nor vice-versa.

The truth of the matter is that science and religion can coexist, they have coexisted, they coexist right now, and they will continue to coexist in the future. They have the same goal of explaining reality—either the material part of reality or the immaterial part of reality. Science and religion are the two "windows" that let humans look at the world they live in. Yet, there is only one world, and it is the same world these two windows try to make known. Closing off either window would make us partially blind.

Science may be advancing much more than religion, but scientific expansion does not necessarily mean that other areas such as religion are shrinking. When it grows, science does not gain territory at all, but it just learns more and more details about its own fixed territory—which is the domain of what can be counted and measured. The rest is not part of its territory but was given away for other "authorities" to manage, including religion.

C.S. Lewis summarized this beautifully when he wrote, "I believe in Christianity as I believe that the sun has risen: not only because I see it, but because by it I see everything else." Religion can open vistas no telescope or microscope can ever reach. When science gives us scientific facts—notice I said facts not hypotheses—we must acknowledge that's the truth, but we also should realize it's only part of the truth. The rest of the truth comes from other facts, notably religious facts. Without the latter part, we get only half-truths at best. Therefore, we must conclude that science and religion not only can live together but also cannot but live together.

For Further Reading

- Barbour, Ian G. *Religion and Science: Historical and Contemporary Issues*. (San Francisco: HarperCollins, 1997.)
- Stephen M. Barr, *Modern Physics and Ancient Faith*. (Notre Dame, IN: University of Notre Dame Press, 2003.)
- Alan F. Chalmers, *What Is This Thing Called Science?* (Hackett Publishing Company, 2013.)
- Francis Collins, *The Language of God*. (New York: Free Press, 2006.)
- Edward Feser, *Five Proofs of the Existence of God*. (San Francisco, CA: Ignatius Press, 2017.)
- Gould, Stephen Jay. *Rocks of Ages: Science and Religion in the Fullness of Life*. (New York: Ballantine Books, 1999.)
- Hannam, James. *God's Philosophers: How the Medieval World Laid the Foundations of Modern Science*. (London: Icon Books, 2009.)
- Jaki, Stanley L., OSB. *The Savior of Science*. (Grand Rapids, MI: Eerdmans, 2000.)
- Polkinghorne, John. *Science and Theology: An Introduction*. (Minneapolis: Fortress Press, 1998.)

Index

A

ad-hoc............ 30, 59, 106
Albert the Great .. 152, 172
annihilation.................193
Aquinas ..20, 74, 100, 113-
   14, 138, 156, 169, 172,
   177, 179, 182, 188, 190-
   91, 198, 205-207, 209
Archimedes 143, 152
Aristotle.....20, 38, 40, 43,
   113, 152, 167, 190, 194
assumptions 140, 155
Atkins, Peter.......... 66, 191
Augros, Michael 21, 75
Augustine 123, 125, 134,
   161, 172-73, 178
Avery, Oswald 62

B

Bacon, Francis..... 6, 153-4
Bacon, Roger...152-4, 162,
   172
Baronius, Cesare 183
Barr, Murray 26
Barr, Stephen 127, 144,
   158, 192
Bastian, Henry 53
Bateson, William... 46, 48,
   108
Benedict XV 175

Benedict XVI172, 174, 208
Berzelius, J.J...............108
Big Bang...... i, 87, 92, 137,
   143, 176, 185-189, 191-
   192, 195-98, 200-202,
   208, 215
Bohm, David................116
Bohr, Niels...... 32, 70, 111,
   116-17, 141
Book of Nature 172-77,
   181-87, 189, 193, 195-
   201, 205, 207-208, 215-
   16
Book of Scripture 172-178,
   179-185, 188, 190, 193,
   195, 197-201, 207-08,
   215-16
Boyle, Robert 154
Brahe, Tycho...........30, 43
Brecht, Bertold 153

C

Carroll, Sean 10
Carroll, William E...... 192,
   197
Casimir, Hendrik..........82
*Catechism of the Catholic
   Church*......109, 169, 178
causality.....6, 117, 146-49,
   155, 195, 214-15

Cesalpino, Andrea.........37
ceteris paribus...............13
chaos 143-46
Chargaff, Erwin.......62-63
Chesterton, G.K.....81, 109
coherence theory......... 114
common descent.... 54, 56
concept-laden..... 24-5, 33,
   105-06, 211
confirmation 27, 30-2, 62,
   64, 150
contingency.................102
Copernicus, Nicolaus.. 30,
   42-3, 168, 177
correlation...... 6, 13-4, 147
Correns, Carl 46
correspondence theory
   ....113-15, 129
creation out of nothing
   ....187-189, 191, 193
Crick, Francis 48, 62-4, 71,
   96, 108, 110

D

Darwin, Charles . 19, 54-6,
   71, 93, 131, 205-07
Davies, Paul C. 73, 158
Dawkins, Richard......... 66
De Vries, Hugo............. 46
deduction 5, 29, 31
Divine Intellect......133-37,
   139, 151, 156-57, 159-60,
164, 166, 213-14
DNA 48, 62-64, 71, 93,
   108, 110, 118
Donohue, Jerry.............63
Duhem, Pierre 162-63
Duhem-Quine thesis 34
Dupré, John..................11

E

Eddington, Arthur 61, 106
Einstein, Albert.... ii, 32-3,
   36, 60-2, 69-70, 85, 87-
   8 111, 114, 116, 141, 143,
   147, 150-51, 158, 171,
   190, 196
empirical cycle... 26-7, 36,
   69, 84, 92, 99, 173, 211
evolution.......54-6, 68-69,
   144, 173, 176-77, 197-
   200, 205-09, 215
experiment 2, 4, 13, 15, 17-
   8, 22, 32-5, 40-1, 45-6,
   53-5, 64, 76, 78, 81, 85,
   91, 99-100, 106, 117,
   119, 136, 149, 151-4,
   164-65, 215

F

Fabricius, Hieronymus.38
falsification ...27, 30-6, 41,
   43-4, 46-7, 52-3, 60, 96,
   106, 112-13, 165-66
Faraday, Michael...85, 112

fermentation .. 52, 107-08, 147-48

Feser, Edward 78

Feyerabend, Paul 77

Feynman, Richard 117

First Cause ...74-5, 82, 90, 99, 101-03, 133-35, 160-61, 184, 188-91, 194-96, 206, 214

Fleming, Alexander...56-8

Flexner, Abraham 85

Franklin, Rosalind 63

Freud, Sigmund 71

G

Galen 37-39

Galileo Galilei.. 29-30, 42-59, 61-2, 111, 153, 162, 172, 177, 183-84

gene .. 22, 26-7, 44-48, 62, 93, 104, 108-09, 111, 118, 122-23, 126, 151

Genesis 1........ 178-79, 200

genetics .. 44-6, 48, 71, 73, 144

geocentrism........... 42, 114

Grand Unified Theory (GUT).............. 9, 81, 88

gravitation...11, 59-60, 72, 74, 88, 128, 130-31, 198

Grosseteste, Robert.....152

H

Haldane, John 148

Hanson, Norwood 8

Hart, David Bentley.... 190

Harvey, William..... 37-40, 52, 87

Hawking, Stephen .66, 73-4, 82, 187-88, 190-91

Heisenberg, Werner ...116, 169

heliocentrism. 176-77, 208

Herschel, Willian and Caroline 59

Higgs boson 194

Hill, James................. 148

Hooke, Robert 107

Hoyle, Fred 186

Hubble constant ...202-03

Hubble, Edwin.. 186, 202-03

Hume, David. 17, 140, 147-49

Husserl, Edmund 97

hypothesis.. 12, 18, 22, 26-34, 36, 39, 46, 47-52, 59, 61, 64, 84, 87, 100, 150, 166, 207, 211

I

induction.... 3-7, 11, 17, 20, 22, 25, 31

inductivism 6, 7-13, 15, 17, 20, 25, 28, 31, 33, 39-40, 95

interpretation.. 20, 44, 85, 92, 95-8, 110, 113, 116-17, 119, 121, 177-78, 199, 212-13

isotopes . 18, 93, 109, 203-04

J

Jaki, Stanley .90, 118, 166-67

Jeans, Sir James ..137, 157

John Paul II 171, 183, 207, 209

K

Kant, Immanuel..........140

Kekulé, August............112

Kircher, Athanasius 40

Koch, Robert 22

Kreeft, Peter ....ii, 194

Kuhn, Thomas... 34-5, 163

L

Lakatos, Imre 35

Lavoisier, Antoine 36, 106, 110

Le Verrier, Urbain... 58-61

Leibniz, Gottfried.. 73, 134

Lemaître, Georges 69, 137, 168, 186

Lewis, C.S..............71, 216

Linnaeus, Carl 45

Lister, Joseph 51, 88

Lorenz, Konrad............67

Lyon, Mary.................26-7

M

Marshall, Barry.............87

Maslow, Abraham 79

Maxwell, James 112

Mayr, Ernst.................140

Medawar, Peter ...28, 100, 113

Mendel, Gregor.. 44-8, 88, 93, 108, 131, 168

Merleau-Ponty, Maurice 115

Mill, John Stuart .. 6-7, 15, 22

Miller, Stanley 53

Minsky, Marvin 71

monotheism.............. 200

Morgan, Thomas Hunt 48, 108

N

Needham, John 41

Newton, Isaac ... 11, 16, 30, 59-2, 72, 88, 108, 110, 131, 167-68

O

Oppenheimer, J. Robert
...............................163

P

paradigm ...35-6, 45-6, 56, 87, 211

Pascal, Blaise 29, 138, 160, 168

Pasteur, Louis ..22, 24, 41, 51-4, 88, 107, 147, 168

Pauli, Wolfgang............ 32

Pauling, Linus.............. 63

Perry, Ralph Barton..... 80

Pius XII 207

Planck, Max....86, 88, 116, 147, 168-69

Plato 23, 97, 134, 189

Polkinghorne, John ... 137, 158, 181

Popper, Karl ... 19, 32, 108, 137

Pouchet, Felix............52-4

Priestley, Joseph . 106, 110

principle of indeterminacy
.... 116

principle of uncertainty
...........................116, 118

Prout, William..............18

Ptolemy 30, 42-3, 114

Putnam, Hillary 112

Q

quantum tunneling....188, 192

R

Rabi, Isidor................109

radioactive dating......203, 205

radioactive decay ..32, 118, 204-05

realism 63, 111-13, 115, 140

Redi, Francesco 40

reductio ad absurdum 101

research program 35-6, 45-6, 54-6, 59, 62, 64, 211

retrograde motion 42-3

Robinson, Arthur.........68

Royal Society of London
.... 83, 153

Russell, Bertrand....16, 31, 70, 167

Ryle, Gilbert.................78

S

Sagan, Carl...... 66, 71, 188

Sandage, Allan............203

Schleiden, Matthias.... 107

Schrödinger, Erwin 72, 80, 116

Schwann, Theodor...... 107

scientism.67-9, 76-83, 90, 177, 212

Semmelweis, Ignaz 49-51, 88
Serveto, Michael 38
similarity 16, 21-3, 105, 107-08, 125
Smolin, Lee 188
Spallanzani, Lazzaro 41
spontaneous generation 39-41, 51-3
Stark, Rodney 163
stellar parallaxes 43
Suárez, Francisco 184
superstring theory 88
Sutton, Walter 48

T

Tarski, Alfred 129
test implications. 18, 26-7, 29, 31, 64, 157, 211
theory-laden 24-5, 33, 105, 211
Thomas Aquinas .... *See* Aquinas
Tjio, Joe 87
tower argument 44
Traube, Isidor 108

U

Ussher, James 199, 201-2, 205

V

vacuum . 29, 111, 188, 192-193
Van Leeuwenhoek, Anthonie 154
variable ... 3-15, 29, 41, 50, 151
verification 31, 33
Vilenkin, Alexander ... 188, 192
Virchow, Rudolf 107
Von Tschermak, Erich .. 46
Vulcan 28, 60, 87, 112, 132

W

Watson, James .. 48, 62-4, 71, 108, 110
Wegener, Alfred 88
Weinberg, Steven 157
Wheeler, John A. 86
Whewell, William . 83, 154
Whitehead, Alfred North 70, 162
Wilkins, Maurice 63
Wilson, E.O. 165
Woods, Thomas E. 163

Z

Zeeman, Pieter 86

Praise for the Book

"Gerard Verschuuren has written an accessible and enjoyable introduction on the nature of the scientific project, the origins of its successes and limitations, and its fundamentally positive relationship with religion."

> —Karin Oberg, Professor of Astronomy at Harvard University, Leader of the Öberg Astrochemistry Group at the Harvard-Smithsonian Center for Astrophysics.

"This book is an excellent introduction to the philosophy of science. As a scientist, Verschuuren has an excellent capacity to describe science's basic methods as well as its limits. The book also shows how philosophy and religion are actually useful in keeping science a grounded discipline. Religion, and philosophy, especially in the Judeo-Christian tradition, actually make science possible in ways that might surprise many scientists. Find out yourself."

> —Fr. Jeffrey Langan, PhD, PhD, Senior Fellow of the Principium Institute, Priest of the Prelature of Opus Dei.

About the Author

Gerard M. Verschuuren is a human geneticist who also earned a doctorate in the philosophy of science. He studied and worked at universities in Europe and the United States. Currently semi-retired, he spends most of his time as a writer, speaker, and consultant on the interface of science and religion, faith and reason.

Some of his most recent books are:

- *Darwin's Philosophical Legacy—The Good and the Not-So-Good.* (Lanham, MD: Lexington Books, 2012).
- *God and Evolution?—Science Meets Faith.* (Boston, MA: Pauline Books, 2012).
- *The Destiny of the Universe—In Pursuit of the Great Unknown.* (St. Paul, MN: Paragon House, 2014).
- *Life's Journey—A Guide from Conception to Growing Up, Growing Old, and Natural Death.* (Kettering, OH: Angelico Press, 2016).
- *Aquinas and Modern Science—A New Synthesis of Faith and Reason.* (Kettering, OH: Angelico Press, 2016).
- *Faith and Reason—The Cradle of Truth.* (St. Louis, MO: En Route Books, 2017).
- *The Eclipse of God—Is Religion on the Way Out?* (St. Louis, MO: En Route Books, 2018).
- *The Myth of an Anti-Science Church—Galileo, Darwin, Teilhard, Hawking, Dawkins.* (Kettering, OH: Angelico Press, 2019).

- *At the Dawn of Humanity—The First Humans.* (Kettering, OH: Angelico Press, 2020).

For more info:
http://en.wikipedia.org/wiki/Gerard_Verschuuren.

He can be contacted at www.where-do-we-come-from.com.

www.ingramcontent.com/pod-product-compliance
Lightning Source LLC
Chambersburg PA
CBHW032223080426
42735CB00008B/691